Perfect Ape to Perfect Idiot?

Neil van Nostrand

Warm regards

Neil

TRAFFORD

USA ▪ Canada ▪ UK ▪ Ireland

Note for Librarians: A cataloguing record for this book is available from Library and
Archives Canada at www.collectionscanada.ca/amicus/index-e.html
ISBN 1-4251-1098-3

Printed in Victoria, BC, Canada. Printed on paper with minimum 30% recycled fibre.
Trafford's print shop runs on "green energy" from solar, wind and other environmentally-friendly
power sources.

TRAFFORD
PUBLISHING™

Offices in Canada, USA, Ireland and UK

Book sales for North America and international:
Trafford Publishing, 6E–2333 Government St.,
Victoria, BC V8T 4P4 CANADA
phone 250 383 6864 (toll-free 1 888 232 4444)
fax 250 383 6804; email to orders@trafford.com
Book sales in Europe:
Trafford Publishing (UK) Limited, 9 Park End Street, 2nd Floor
Oxford, UK OX1 1HH UNITED KINGDOM
phone 44 (0)1865 722 113 (local rate 0845 230 9601)
facsimile 44 (0)1865 722 868; info.uk@trafford.com
Order online at:
trafford.com/06-2857

10 9 8 7 6 5 4 3 2

This book is dedicated to all our sons and daughters,
their children and to their children's children.

Acknowledgements

As a computer-illiterate, I am very grateful to Matthew Slade for key-boarding my many drafts of this story over a five-year period. Master printers Dan Sweeney, Print Shop and Deborah Seary, Copy Centre were likewise of inestimable help in getting the manuscript into book form.

Good friend and wildlife mentor Dr. Donald G. Dodds provided valuable insights. Paul Tufts and Mike O'Brien, two of Don's Masters students at Acadia University, followed up on my earlier beaver work to round out Nova Scotia's classic boom-bust beaver population study.

Gary Saunders, author of 10 books and veteran editor, devoted many hours to the various versions. A simple "thank you Gary" is putting it mildly.

Thanks also are due to many unnamed Nova Scotians I've worked with during 25 years as a wildlife biologist.

I was particularly fortunate (beyond words) to have been part of Nova Scotia's first team of four wildlife biologists in Clarence Mason's new Wildlife Division. Don Dodds, Merrill Prime and Fred Payne, each in our own specialty, couldn't have been more compatible or more productive in our chosen profession.

In my furbearer specialty, I seemed to have been born

to do research in the Tobeatic Sanctuary (now Tobeatic Wildlife Management Area). Charlie Jones (Sanctuary Warden), contributed his indispensable woods-wisdom as a long-time Sport Guide, to perfectly complement this long-term furbearer research and make my days in the Tobeatic so memorable and rewarding.

Above all, I thank my spouse Erica Garrett for her on-going astute advice and guidance from the start – truly a collaborative adventure in publishing.

Credits

Covers

Rob van Nostrand (www.PerfectPhoto.ca.rob@ PerfectPhoto.CA) photographed the painting and sculpture, then designed the composite front cover.

The painting and sculpture were done by the author in 1989 as part of the Rudolf Steiner Waldorf teacher training at Hungry Hollow, N.Y., U.S.A.

Other Photo Credits

p. 14 Horace Rand farm - Richard and Jeanita Rand

p. 56 Wolves - Andrew Kiss in *Cool Woods*

p. 68 Hillary House - Sheila Richardson

Glossary of Biological Meanings and Vital Concepts of Key Words

Acquire an additional trait or ability gotten by unspecified means.

Aggressive marked by driving force.

Biology study of life including interaction of life systems.

Carrying Capacity population of living organisms that can be supported in a given space at any one time: those not supported are surplus.

Celtic those with acquired Celt tactics mimicked ("monkey see monkey do").

Ecosystem a community of living and non-living things functioning as a unit in Nature.

Environment that which surrounds an organism and interacts with it.

Environmental Crisis collapse of one or more interconnected ecosystems.

Evolution gradual change from one life form into a different life form and the theory for that change.

Fact Photograph of, or information as having actual reality, with or without having personally seen it. For example:
-   **Earth** floating in space;
-   **a nuclear bomb** explosion;
-   **extinctions** of species;
-   **degradation** of living ecosystems.

Global Warming CO_2 and other accumulating "greenhouse" gasses that trap infrared radiation from the sun causing a global rise in temperature.

Holistic *(1926) biologically speaking, relating to whole complete systems rather than dissection into parts. Thus humanity and environment are a single holistic system.

Imprinting a rapid juvenile learning process that permanently modifies adult behaviour.

Introduce brought into favourable habitat for the "first time".

Lost lacking assurance and self-confidence, deep helplessness, disorientation from known markers.

Nature all existing plant and animal life outside of humanity.

Non-compostable Garbage, Waste, Refuse discarded packaging and things made obsolete by new improved or different things.

Oil Well *(1847) the first source of crude oil energy, now deemed essential to the existence of developed civilizations.

Pollution contaminants harmfuly affecting the environment and living organisms.

Population density the concentration of individuals per unit of living area.

Primary producer (human) one directly involved in exploiting a natural resource to satisfy basic needs such as food, fur, fiber, etc.

Sustainable capable of performing undiminished in perpetuity all things being equal as to carrying capacity, environment, social, health, etc.

Territory a defended geographic area.

Wealth all material objects having economic human value in existence at one time.

Wildlife non-domesticated life, wild by nature.

*Date first used

Contents

Introduction

This story begins with some memories of growing up on our family farm in Ontario in the 1940's. Wildlife interested me from an early age. I began to marvel at how every species can thrive so perfectly and easily compared to our complicated effort to live by farming.

Graduating from the Ontario Agricultural College with a BSA in agronomy and an MSA in wildlife, I was lucky to find a full-time permanent career doing field work researching ways to better manage pheasants and beavers in Nova Scotia.

Now imagine, if an alien, landing on Earth, had asked, "Why do humans seem so hell-bent on destroying each other and undermining their own life-support systems, some even threatening to blow up the only living planet they know"? This would not have struck me as an unreasonable question.

Surprisingly, this actually seemed to be the question I was drawn to since I discovered how every other species cooperates ingeniously to build up the life on Earth, creating an indescribable beauty humans inherited. Why are we so focused on all this destruction? This led to some unexpected and interesting objective speculation, with much food for some thoughtful changes for the better.

Being a farmer at heart, and seeing the unsustainable

direction of our food production, I wrote an article published in Agrologist, Winter 1975. However, farmers had been convinced to make farming a business, no longer the way of life I know. So it was—food production for profit to feed increasingly hungry people in the world. Ironically there are many more millions starving than ever, and, farming has become a very risky business. As a professional, practicing, practical, wildlife biologist, I decided to look into our human quandary. The result is this dissertation and its conclusions; some uniquely different.

The human population is booming globally for the very first time. But, we are on the cusp of an inevitable bust. Past busts have always had escape routes—people could always move to a new pristine area and repeat the boom as have many people immigrating to Canada. I've used my biological observations on beaver populations to illustrate the natural characteristics of both booms and bust, and why busts must inevitably follow booms.

With this story and my conclusions, I would hope individuals could better understand what's ahead in the wake of our current phase of massive destruction of the Earth's natural ecosystems, and why this affects all life globally. Possibly individual readers will be able to make wiser choices for a happier and healthier life with a better understanding of our global plight

An Overview

_____

We "Celts" (ie: people in Celt-led nations) see and know all; but we constantly need reminding. We each try to do our best, looking for approval in his or her solitude. We have everything and nothing. Our deep malaise about not "fitting" is authentic at every level, including "in Nature". In one sense, we have left the womb, and our jungle home, and keep trying to find a kindred soul to console and be consoled by in our human-made jungle. In another sense, we are "lost".

Our computer screen has access to <u>all</u> the vital pieces of the "picture puzzle" of humanity's surviving on this planet. British journalist Elaine Morgan, in her very persuasive _"The Aquatic Ape Theory"_, provides the quintessential example of how to stitch together <u>logically</u> the known pieces of human evolution, to recreate the whole beautiful living quilt of our existence.

In _"Perfect Ape"_ I've used essentially similar logic to stitch together some known facts and relevant truths to portray blow-by-blow as it were, humanity's imminent biological disaster.

By my own "idiocy" I've tried to demonstrate how we got lost, where we are now and why we need a "town meeting" to decide how to proceed, each in our own life. Pretending we have no choice is, to me, our worst-case scenario. "Nothing is more terrible than to see ignorance in action" Goethe, 1749-1842.

Let's look at a few examples.

MP Joe Jordon (Prescott, Ont.) recently introduced to Parliament as alternative to GNP, the concept of GPI—Genuine Progress Index. Essentially it is an annual report outlining

a)   Major Canadian measurable goals

b)   Evidence of achieving each (or not)

c)   Suggested changes and improvements.

Medicare is our current example. Others may be water, energy, poverty, defence, NAFTA, "gun control" and so on.

Each of us needs periodic reviews of his/her personal life situation too. Are you very satisfied ❑, satisfied ❑, dissatisfied ❑, or very dissatisfied ❑ with life? "What needs changing and what's my plan for achieving change?" (See Joy Brown's *The Nine Fantasies That Will Ruin Your Life and the Eight Realities That Will Save You*, as an example.)

Change is a Celt's middle name. All we need is motivation. Motivations among Canadians are as diverse as our more than 100,000 named "causes" (Canada has more than 78,000 "Charitable" causes alone.)

Developed nations in particular, focused on "terrorism" after September 11, 2001, when New York's World Trade Centre Towers were destroyed within minutes of each other by suicide pilots crashing two loaded passenger planes into them. More than a year later most people were still very muddled about what motivated this action and what the response should be.

Once Louis Pasteur proved that tiny germs were the cause of the horrendous plagues of the Middle Ages, science made a quantum leap. They replaced "God-based" explanations for diseases (churches were filled with the dead and dying) with the belief that if Nature could be fully understood, people would be more in control of their own well being.

In the last 150 years in fact, science has exercised a degree of control over Nature hitherto unimaginable, and <u>did</u> control diseases through our understanding. By implementing massive changes to water and sewage systems for instance, we did largely control several lethal diseases. Similarly, science revolutionized both food production and pest control using fossil fuels and chemicals. "Microbiology and meteorology now explain what only a few centuries ago we considered sufficient cause to burn (8,000?) women to death." C. Sagan 1996.

But each breakthrough has it's own cost. Globally it is now crystal clear that the wealth of developed nations has been amassed by robbing Nature's capital storehouse; that science-dominated Celt-led nations are totally addicted to massive consumption of energy, especially oil, the real "goose" that has been laying all our "golden eggs", and that if we continue at this rate, our goose will quite literally be cooked.

Witness a few biological facts:

Oil – Our fossil fuel river of life – four tonnes of oil-equivalents of energy per man, woman, and child every year in developed nations. (Statistics Canada, 2005)

Soil – Soil nutrients (created by Nature over 10,000 years) are being "mined", polluted and paved over. Moreover, usable soil nutrients in our feces are contaminated by chemicals yet are being flushed or dumped into fresh and salt waters.

Health – Degenerating health has links to diet, lifestyle and environment. Current threats include radioactivity, pesticides and pharmaceuticals.

By our attempt to live proudly outside Nature, we have created in a single generation (mine), a unique biological anomaly in that we Celt-led nations are:

- out of touch with the Nature-system supporting all of us,

- cut off from any meaningful communication with that system (we chose to cut ourselves off) and,

- in the euphoric" free-fall" mental state of a dependent three-year-old.

The longer we delay "biting the bullet" by collectively, honestly sitting down and facing the gravity of our situation, the lower are the fast-shrinking odds of our ever recovering a niche in Nature. Indeed the beautiful Celt no longer fits any niche in Nature. Moreover we are taking down, many, many species with us. And, as in the well-documented dinosaur "extinction event" 65 million years ago, Gaia will again be set back to an earlier succession. How far the back swing will go depends on our capability to change. Even a non-choice is a choice.

Part I:

My Observations
_____

Growing Up and Adult Experiences

Have you noticed how we are destroying the earth? Have you ever wondered why we keep destroying Nature? I've wondered since boyhood about these things. So I've decided to put down on paper some of my memories, observations and conclusions. Looking back, I loved boyhood on our family farm in Vandorf, Ontario, though at the time it was hard work. And while I was extremely fascinated with all animals, including wildlife, I always planned to farm.

Dad got back from the War, and one day took a carload of us boys to the neighbour's tree planting project. He wanted us to see first-hand the value of planting trees on land where the original white pine had been cut down, as indeed they had been on our farm too.

On the neighbour's farm I vividly remember seeing a big area with giant old pine stumps standing two feet off the ground on their spindly roots. It was explained that once these great trees could no longer protect the soil from wind erosion, the loose sandy soil that produced them began to blow away, drifting like snow onto the next field, and the next.

Once this happened, the only way to stop further erosion was to plant thousands of 25 mm (10") high fast-growing evergreen seedlings of jack, red and Scotch pine. We were shown the results, beautiful green Christmas trees, planted only six or so years earlier. So every spring after that, we boys planted hundreds of trees. I still love planting trees; it's addictive.

When I got old enough to help our three hired men with farm chores, I did so before and after school, and cleaned out the pigpens every Saturday. It was natural. We had eleven milk cows, hand-milked twice a day. The milk and cream were "separated" using a hand-cranked standard centrifugal separator. The resulting cream was sold for butter, and the skim milk fed to the young pigs and calves. Every morning in summer my special job was to brush down the five big Clyde and Percheron workhorses. They were used year-round, hauling not only all the farm implements, each in its own season, but manure, hay and firewood.

I remember seeing our first tractor. Its uses were discussed with a great many doubts, even down to the rubber tires. (Prairie farmers used all-steel wheels). It was conceded that we probably could shorten the implement hitches for the heaviest work of plowing, and likewise on the four-horse binder. But nobody could imagine doing without horses in mud or snow; or for hauling a ton of hay. As for the animals' uncanny ability to move from haycock to haycock with only the slightest "chirp" (like a kissing sound) from the teamster, and of stopping on "whoa", no machine could beat that. Horses would always be needed. To me they were pure magic.

Our farmhouse electric lights were a real luxury when they first came, but we always kept house lamps for emergencies. In the barn we kept using kerosene lanterns.

A large windmill pumped water from a 100 meter deep well by the barn and into a cement house cistern. Since the house cellar was higher than the barn cellar, the water then flowed by gravity into the barn taps. But all the house water came from a hand pump in the kitchen, as needed. A big tin dipper hung by the pump for drinking out of, and for carrying water to the stove for cooking.

There was no indoor bathroom. Everyone used the outhouse in the back part of the woodhouse next to the house. Baths were taken every Saturday night before bedtime. Placing the big round metal tub next to the kitchen stove, Mum filled it with nice warm water. First my baby sister, then all five of us boys had a good scrub with grandma's homemade lye soap.

My summers during the War were spent working on the farm and tending our Victory Garden. Victory Gardens were a way of coping with wartime scarcity of food; urban backyards had theirs too. Grandma had a nice flock of Barred Rock hens. These provided lots of fresh eggs in the spring. Extra eggs were stored in big crocks in an airtight liquid "water glass" which preserved them for months when the chickens were not laying. Without modern high protein feed, hens stopped laying about July.

Grandma often planned on having chicken for Sunday dinner. The night before, we would pick out one of the old hens that she knew wasn't laying any more. Holding the chicken's feet in her left hand, she would wring its neck with her right by a simple twist of the bird's head. Then all the feathers were picked off and the bird hung by the feet to cool overnight in the cellar. Next morning the guts had to be skilfully removed so as not to contaminate the meat. The neck, liver and gizzard were separated and cleaned to make giblet gravy. Grandma taught me these basic steps I still use when killing and dressing game birds.

After the Great Depression, things slowly began to improve. "Tramps" riding the tops of freight trains no longer came to our door looking for a meal in return for doing odd jobs. In the late '30s a new electric water pump was installed in the basement to provide hot and cold running water from taps upstairs and down. The water was heated in a waterfront iron box bolted to the side of the woodstove, replacing the old "reservoir" there. A hot water tank sat next to but higher than the waterfront, causing water to circulate by convection into it. At the turn of a hot water tap, cold water entered the bottom of the tank, forcing the hot water at the top, into the pipes above.

Having hot and cold water under pressure meant we could have a bathroom upstairs complete with a flush toilet—all luxuries to help my mother handle six growing kids. She also looked after old Aunt Edith, Dad's mother's sister who had a lame knee since childhood.

In time we also got a new shiny white refrigerator to

replace the icebox in the porch. Ice for this came from the icehouse beside the house. Insulated with sawdust, it preserved the huge blocks of ice all summer. The 'frig' saved two men two weeks' work sawing blocks of ice at the lake in winter and packing them into the icehouse.

At the age of 20, much as I dreaded the idea of more schooling, I decided that if I was going to farm, a couple of years at the Ontario Agricultural College in Guelph, Ontario wouldn't hurt. This tenuous first step led to a four year B.S.A. specializing in field crops, followed by two more years towards a Master's degree in Wildlife Management there.

Then came marriage, a child, and a wildlife job in Nova Scotia where I spent twenty-five years doing furbearer management research throughout this wonderful province.

After compiling an educational manual for first-time trappers, I retired from the civil service in 1986. Later my spouse Erica and I took up mixed organic farming for nine years before selling the farm, lock stock and barrel, to a young family to carry on with certified organic vegetable production. Retired in Wolfville, Erica and I decided to combine our talents to record this story, and what I learned, beginning with my arrival in Nova Scotia.

Farm Transformation

Good agriculture land was cheap in Nova Scotia when I began looking for a place in the country in the '60s. I could have bought a whole farm for $10,000, but with my full-time wildlife job, leaving scant time for farming, it seemed smarter to settle on two hectares (5 ½ acres) with a typical big old farmhouse and barn on Starrs Point. It cost $4,250—exactly the same as my annual salary. Our young family enjoyed being only five miles from town with wonderful neighbours nearby. I loved the many relics identical to those I grew up with.

Our 80-year-old neighbour, Horace Rand, happened to be one of the first farmers I interviewed while researching farm history in 1958. Horace plowed a patch of garden for us with the last team of workhorses on the Point. He was so proud of them. But then they had to be put out to pasture and they were rarely used again.

Horace Rand Farm, Starrs Point 1960

Little did I realize how fast things were changing. Horace's grandsons, graduates of the Nova Scotia Agricultural Collage in Truro, were already planning big things. One of these was to grow Nova Scotia's first tobacco crop. This meant clearing old messy fencerows to make huge fields, and getting rid of old apple trees. The once famous Annapolis Valley apple orchards still dotted many farmscapes.

Everything would be totally mechanized. No longer would dad have to work off the farm to pay the mortgage. The whole family could spend their winters in Florida. This was the farm vision that drove young farmers to toil long hard hours every summer. In one generation these modern agricultural teachings completely transformed the entire Annapolis Valley. They were the same things I was taught. They are still taught in every developed country, and are probably encouraged or subsidized as solutions in hungry Third World countries.

Beginning in the eastern end of the Valley, what used to be hundred-acre, self-sustaining family farms were transformed into corporate family farms of thousands of acres. One hundred-acre fields of single crops like corn, soybeans, carrots, potatoes, peas, broccoli and onions became commonplace. While small by California standards, they were and are massive by Maritime standards.

But such corporate family farms are always at the mercy of the whims of global prices, government subsidies and bureaucratic safety nets', not to mention recent devastating droughts. In time some bankruptcy was and is inevitable. The surviving state-of-the-art operations, specializing in horticulture (like broccoli), chicken or pig are urged by government and banks to get still bigger, more mechanized, more efficient and presumably more globally competitive. Meanwhile, our bureaucrats argue in international courts about who is 'dumping' or unfairly subsidizing the cheap food consumers enjoy. We, in turn are often unaware of the near bankrupt state of many of our local farmers who look so prosperous.

The traditional wisdom and love of the older farmers, which used to sustain not only the soil but also the whole community, are gone. Upcoming farm generations facing financial ruin can hardly be blamed if they leave for a better life in town. The farm illusion—that so few can produce so much so cheaply, is believed by more than just unknowing, uninformed consumers. In fact it is taken as a sign of success. What a sad parody! It now takes up to 20 calories of oil energy for us to produce one calorie of food consumed. Meanwhile market pressures are dictating even more oil consumption.

"Health Canada"

Thanks to speedy transport, efficient refrigeration, plant and animal breeding and modern communication to name a few innovations, most Canadians can now enjoy the same global foods year-round from coast to coast to coast, even in some remote Arctic villages regularly serviced from the south. On the face of it, some of our staple foods appear to have a familiar look and taste. Certainly foods produced and delivered to the farm gate—milk, cows, pigs, chickens, eggs, fruits, grains and vegetables—look the same as ever. In fact, however, our diet has been transformed even more radically than our farms.

Medical and scientific studies now implicate our diet and

lifestyle as major factors in our increasingly serious health problems, problems which occur at ever-earlier ages. Over half of us, including children, are over-weight, a first sign of trouble ahead. Native people suffer the most. Not only has their traditional life-style been destroyed in one or two generations, but many in the north are now coping with the effects of immoderate consumption of our sweet food and drink. Diabetes is rampant, afflicting 25% or more of our Native population. Marginalized and robbed of dignity, they have the highest suicide rates in Canada.

Moreover, today's northern caribou herds carry ten times the allowable limit of residual persistent pesticides, even though these are used only on agricultural crops many hundreds of miles to the south. It's ironic that native mothers are often flown south to have their babies born in a hospital, yet their milk is often too toxic in some to allow them to nurse.

Even with Canada having arguably one of the best health care systems in the world, Canadians probably have one of the highest percentages of degenerative afflictions. This could be because ours is one of the coldest countries, which means we use the most resources per capita, driving around or staying indoors in controlled environments more so than our counterparts in warmer climates. This sedentary lifestyle encourages us to consume more processed fast food and drink, which undoubtedly aggravate the situation.

Contrasting Food Systems

In Third World countries most farmers still live and work on the land. When I was growing up in southern Ontario during WWII, most people did the same. Today, in stark contrast, Canadians pride themselves in having less than 2% of our population "engaged in farming". At the same time we boast that one in eight or ten people in the work force depends directly on our agri-food industry for their living.

In other words, those not owning or working on the land are involved in farm machinery, fuel, transportation, processing, fertilizer, pesticides, packaging, advertising, selling, research and teaching, to name a few. Each of these sectors accounts for a major share of the consumer's food dollar. With so many involved, the farmer's share may be only two to five cents. And no cost accounting has ever been made for agricultural pollution of the water we depend on, of the air we breathe or of the topsoil lost to erosion .

For instance, approximately two bushels of topsoil are lost for every bushel of corn or potatoes grown. However, Canadians need spend only an estimated 10% of their take-home pay on food. Third World people spend 50% or more of their income and effort on food, since most still produce their own.

Sustainability of First and Third Worlds Compared

Both the agri-business vision and the reality in these two very different worlds confuse the sustainability issue. But at least Third World people who live on the land producing food and fibre have a much more realistic concept of sustainability, in that their daily lives are physically entwined with Nature. Tightly bound to the land, they know first hand what the land will bear and that ill-treating it exacts immediate reprisals. Many have used these same lands for 40 centuries or longer.

By contrast, Canadians who have been using chemical fertilizers and pesticides on soils for only four decades, yet find many of their soils already almost lifeless. Devoid of most of their original organic matter, trace minerals and other nutrients, they can only maintain production with ever increasing inputs of fertilizers, pesticides and machinery. Even then, these crops lack their capacity to sustain the health of livestock and humans as they once did (500 bushels of wheat minerals were in 20 bushels in 1949).

In industrialized nations like Canada, 90% of the citizens have no real contact with Nature outside of golf, bird feeders or at best, gardens so they lack a realistic understanding of their own food system, the "umbilical cord" without which they would die. To them sustainability has to do with business. They fail to recognize that conventional and even organic farming methods constitute an energy sink costing far more calories of fuel oil per unit than are ever produced in food calories. As mentioned, estimates have been as high as 20 calories for each calorie consumed. Even so, densely populated Germany is now subsidizing organic agriculture as their cheapest route to sustainable, secure, high quality water, food and soil.

Farmers of Forty Centuries by F. H. King (1911) is the classic text of permanent agriculture in China, Korea and Japan. It explains the wisdom of oriental practices of maintaining soil productivity indefinitely without the use of any artificial chemical fertilizers, which are so clearly at the root of soil destruction, erosion and depleted health-giving qualities for domestic animals and humans dependent on them.

The wealthy G-7 Celt-led nation farmers have only been able to maintain farm productivity over the past 40 years by large applications of mostly fossil-fuel-produced artificial N P K fertilizers.

In the preface of King's book, L. H. Bailey explains the fallacy of North Americans exporting food and food production technology when both depend totally on very costly and unsustainable inputs of non-renewable energy and vast tracts of farmland. He writes;

> If we are to assemble all the forces and agencies that make for the final conquest of the planet we must assuredly know how it is that all the peoples in all the places have met the problem of producing their sustenance out of the soil….

> We in North America are wont to think that we may instruct all the world in agriculture, because our agricultural wealth is greater and our exports to less favoured peoples have been heavy; but this wealth is great because our soil is fertile and new, and in larger acreage for every person. We have really only begun to farm well. The first condition of farming is to maintain fertility. This

condition the oriental people have met, and they have solved it in their way. We may never adapt particular methods, but we can profit vastly by their experience. With the increase in personal wants in recent time, the newer countries may never reach such density of population as have Japan and China; but we must nevertheless learn the first lesson in the conservation of natural resources, which are the resources of the land. This is the message that Professor King brought home from the East.

Canada's Growing Biological Threats

Canada's prosperity comes mainly from exporting our natural resources like fish, wood, oil and other minerals. Over 80% of all our exports go to the U.S. Naturally the U.S. would, if need be, defend Canada. They need those resources—especially fresh water, not yet being exported (2004). But already on the drawing board are trillion-dollar projects to divert Canada's fresh water south. Water is our last abundant renewable natural resource. In the west it would go via the Rocky Mountain trench. In the east, Hudson Bay would become a giant collection basin for freshwater to be routed to the U.S. via the Great Lakes and the Mississippi watershed.

Canada already has precious little readily available unpolluted fresh water. What there is, is increasingly sold in grocery stores everywhere for about double the price of fuel at the pumps. It's not economically practical to purify polluted water as a human beverage, at least not without adversely altering its natural minerals and other

beneficial qualities. All urban water must by law be dis-
infected—even if it is perfectly pure at the source—usually by adding chlorine.

It is calculated that, if all the planet's six billion people
were to live as recklessly as Canadians do, gobbling up
far more then our fair share of the earth's annual resource
output, Nature would need 4.5 to six earth-size planets
to cope. Even if we all did our bit to help out, such as
buying locally, living in small, energy-efficient housing,
walking, biking and so on, the impact would be minor
since industrialized countries make up only 20% of the
world's population. This 20% annually consumes 80% of
the marketed goods & services produced. Remember,
our per capita energy consumption in oil-equivalents
now averages more the four tonnes per year for every
man, woman, and child. We are told that this energy
consumption must be jacked up each year in order to
avert global economic disaster – that is, a worldwide
depression causing chaotic disruptions.

Part II:

Evolution Theories

Overview

Have you ever wondered if Canada's (and the developed world's) prodigal way of life (see Part I) could be made more sustainable if we had a better understanding of our past millions of years on the planet? That we might set realistic long-term goals for Canada? Part II was written with this in mind after I had read and digested Elaine Morgan's *The Aquatic Ape Hypothesis*.

Before discussing her conclusions, let me state four basic truths on which I base Part II, namely

1. That in our two million year history (some sources estimate 5-6 million) humans have endured many and vast changes, as archaeological records show.

2. That humans, being mammals, are subject to (not masters of) the laws of Nature.

3. That my farm background, my professional wild-life research and my long experience with anglers, trappers and hunters (moderns with strong ties to

Nature), qualify me to understand our human journey with less bias than most.

4. That such a tough-minded, unbiased perspective will help us recognize the folly of our short-term logic and our counter-productive behaviour.

If nothing else, humans are adaptable. We have been change artists throughout our existence. If we must do an about-face, we can. "Where there's a will there's a way". But first we must be convinced of the urgency of the crisis. Ms. Morgan's research into the relevant literature, provides important insight into our adaptations thus far.

A New Human Evolution Theory

Every culture seems to have a story of how people "originated". After Darwin's *The Origin of Species* (1859) and *The Descent of Man* (1871), scientists began looking in earnest for fossil evidence of the "missing link. Many fossils, books and theories later, we are left with very contradictory and unconvincing explanations of the observed differences between our nearest relatives, the apes and humans. Many differences—our large brains, "hairless" body, upright posture, sub-skin layer of fat, ease of swimming from birth, diving reflex, salty tears and so on, are simply unaccounted for.

It often happens that quantum leaps in scientific research and advancement come by surprise, and from people outside the discipline. Elaine Morgan's aquatic ape theory may be such a leap.

In fairness, it was a highly respected British oceanographer Sir Alister Hardy, who first broached this theory in his article *Was man more aquatic in the past?* published in *New Science*, March 17, 1960. He based his theory on observations made over a 20-year period on the many likenesses between aquatic sea mammals and humans. Surprisingly, his article was given wide publicity in the daily press. Elaine Morgan, an award-winning Welsh television programmer, did her own literature research into the new theory and in 1972 came out with *The Descent of Woman*, an instant international best seller. This was followed in 1982 with *The Aquatic Ape*; in 1990 by *The Scars of Evolution*; in 1995 by *The Descent of the Child*; and in 1997 by *The Aquatic Ape Hypothesis*.

The Aquatic Ape Theory of Human Origins

Morgan's *The Aquatic Ape Hypothesis* begins with our progenitors, or proto humans, as perfectly ordinary apes marooned on "Danakil", a very large island that existed five to ten million years ago but which is now under the Arabian Sea.

At that time there began a very long slow drought over this entire sub-tropical region. The resultant slow drying of Danakil's jungles forced a stranded population of apes to look for additional food along the nearby seashore. Here they found an abundance of small sea creatures, which to them resembled the small grubs they often found and ate to supplement their vegetarian diet.

Over many generations, they came to relish these salty

protein supplements. Over millennia, seafood became a vital part of their diet. Cracking the shells of arthropod and clams with stones, as sea otters do, likely became routine for them. Thus, incredibly slow change from tree to sea is postulated for our ancestors on Danakil Island. Scientists estimate that something like 50,000 years of isolation may be needed for even a sub-specific minor physical change to become entrenched. Those apes had perhaps millions of years.

Nature's Evolutionary Stable Strategy (E.S.S.)

Scientists now know a great deal about general evolution. Dawkins' *Selfish Gene* documents Nature's strategy of maintaining with iron determination, the immutable laws governing the stability of all life. Violators suffer Nature's death penalty—extinction. Within the overall strategy of stability, organisms tend to change irreversibly, from the simpler to the more complex state.

Proto-humans, unlike every other species, failed to develop inherited instincts to ensure a stable strategy in their evolution. In fact, from the beginning, our humanoid instincts all inclined against an E.S.S., an anomaly explained in a later section.

The invention of the pebble tool, first to crack shells and later to be fitted with a handle to become an axe or an arrow, is often regarded as the earliest invention of proto-humans. Few such tools are ever found because in Nature they quickly separate into their two components, a stone and a stick. The pebble tool idea was our first "atomic bomb"; its ripples, beginning millions of years ago, are now tidal waves. The idea was probably first used for obtaining prey too big or too elusive to catch using only ape hands or teeth. Skillful use of this tool to obtain large prey soon made the use of larger pebble tools a decisive factor in defending territory. A major improvement in the original idea was gunpowder, which, as one pundit quipped, "is a substance used by civilized societies to prevent the re-occurrence of problems."

The Evolution of "Nature-Destruction"

1. Perspective

Nature, of course, can never be "destroyed": it can only be changed from one form to another. Nature's changes are beautifully organized, have a stable strategy and employ long-term, progressive, building processes. In stark contrast, human–caused changes to Nature tend to be destructive, unstable, and short-term. Before wealthy G-7 countries, in the last 50 to 100 years, began to use such massive amounts of artificial energy, Nature seemed quite capable of washing away our "castles in the sand" with her tides of time.

The example of a natural parasite is instructive. In the midst of "plenty", a parasite such as a roundworm in the gut of an animal, evolved neoteny. Neotony allows them to reproduce at the juvenile larval (worm) stage, forsaking their free-living "adult" form altogether. Celts by contrast have an adult body to reproduce with, but have nonetheless created an immature irresponsible life-style in one generation. Celt-led nations behave as though we can continue to squander, pollute and destroy natural and human resources indefinitely. Our ancestors did the same by killing, stealing, lying, cheating and deceiving, but they had better excuses for doing so. Can we honestly face what we now know?

2. Gaia

Entropy means the diffusion and eventual decline of energy (heat and light), from any source. Basic energy sources, like our sun, millions of years old, generate heat and light via nuclear reactions.

"Gaia" was the Greek "Goddess of the Earth". In fact, scientists have proven that Planet Earth does respond to all external forces, not unlike a living creature. Gaia's perfect, interactive system includes all life (Nature) as well as non-life (water, minerals, etc.). Gaian successions have incorporated the planet's slow cooling, the beginning of life, and such mega-effects as accidental collisions with things like asteroids, one of which may have wiped out dinosaurs 65 million years ago. Gaian life systems are designed so that the chlorophyll of green plants captures the Sun's energy and converts it into sugars and starches. These fuel not only the plant's life processes, but "trickles down" to run <u>most</u> of the other life forces on Earth.

We might say that all green life concentrates the Sun's diffusing energy while animal life disperses it. A third component namely fungi and bacteria complete the loop by decomposing dead plants and animals into their original components.

Modern humans extend the process exponentially. We

burn up massive amounts of non-renewable coal, oil and gas. In fact, we have become so accustomed to using increased quantities of this type of stored energy, millions of year old, that if the world's production of oil alone were to decrease significantly (as it has to sooner or later), a global depression would ensue. No wonder Kyoto's proposals to cut back to 1990 levels frighten wealthy nations.

3. Pebble Tool to Extinctions

According to the aquatic ape hypothesis, proto-human apes on Danakil Island two plus million years ago were slowly and step-by-step forced by unique circumstances and isolation, to change radically in order to survive. Thus, to succeed at catching and eating seafood, this previously vegetarian ape underwent many subtle evolutionary modifications in physiology and behaviour over these millions of years. Their eventual adaptation to eating a wide diversity of food—even after they left Danakil for Africa's steppes, is very evident in humans today. It is one reason why we are so out of balance with Nature and with all other species. To sustain this "habit", we created artificial human and other habitats such as fish farms. These developments are now the main cause of rapidly rising rates of species extinctions.

One could say that our destructive habit began with learning to hunt and kill small creatures along the seashore of Danakil Island. Over millions of years, hunting and killing also became very important for human survival. This legacy is still central to our current Celt concepts. Moderns are muddled meddlers beset by many mixed feelings about our medicines, defences, security, food, GMOs, sport hunting and the like.

Extinction of species is not new. Scientists now see extinction as being a normal, if not essential, process in Gaian evolution. Our present number of living species on earth may approximate 50 million—many of them still unnamed. Yet scientists tell us these may make up as little as five per cent of all the species that have <u>ever</u> existed on Earth. However, our current rate of species extinction on earth is about one every day and increasing. To palaeontologist evolutionists this looks like the beginning of another "extinction event" on a scale similar to the one that wiped out all dinosaurs. What makes our "event" unique is that it is being caused by human development.

It makes one wonder how far we wealthy nations can or will lead in our present drive to develop the other 80 % of our six billion population—especially when it is motivated by gain as much as by altruism. Meanwhile, 800 million go to bed hungry every night, and millions are said to die from starvation weekly, the bulk of them in war-torn parts of Africa. [CBC News]

4. Evolution of Human "Pests"

A. Background

Throughout human and wildlife evolution, all diseases were parasite or germ-oriented ("germs" are one-celled organisms or virus parasites) that function as two of Nature's stabilizers. They prevent any species from out-stripping its food supply. Each species has its own parasites, evolved to prevent over-population. As a rule parasites are quite host-specific, and do not eliminate their hosts, which of course would mean their own demise.

The term "pests" includes parasites and diseases of both humans, animals and plants. It is also used to include anything humans don't happen to like for any reason. Commonly this includes weeds, crop destroyers or even just creepy crawly things, spiders etc.

I've used the pest concept to include "introduced species" using examples of the Varoa honey bee mite, threatening apiculture, the eastern coyote booming out of control and the beavers of Nova Scotia, once a pest. I've used the beaver boom-busts I studied to illustrate the natural characteristics of each and why busts always follow booms.

Proto-humans got rid of land parasites like head lice, body lice, scabies, mites and such, when they began going into salt water in search of food. The sea also gave our ancestors some protection from land predators – until they returned inland. In time, however, success in food gathering led to population increases in the Middle East "fertile triangle" which led to hunting and gathering scarcities. About 10,000 years ago, the domestication of plants and animals began in earnest. However, when domesticated animals were brought into intimate contact with humans, certain of the animals' disease organisms crossed over to humans (zoonosis) and in time became endemic in humans (e.g. TB and brucellosis from cattle).

Eventually these germs evolved the ability to spread directly from human to human. In the Middle Ages these outbreaks reached epidemic proportions. Plagues like the "black death" wiped out half of Europe's population. Gradually, eventually human survivors developed natural immunities. This, along with increased sanitary awareness, gradually reduced the toll from these new parasite pests.

In those days, one's fate, good or bad, was attributed to pleasing or displeasing the whims of God. Once Pasteur "proved" that germs, not God, cause diseases, scientists began to believe that once Nature was understood, man would be in control.

B. Science

a) Immunity—"having a high degree of resistance to..."

The immunity principle is illustrated by our modern inoculation of children with a tiny bit of a germ disease's dead or weakened body, which stimulate the child's own antibodies to prepare its system in case the real "bug" comes along. This principle* has worked miracles in ridding modern humans of many of our most devastating microbial diseases. Among the last of these was polio whose crippling effects where defeated in 1952 by Jonas Salk's vaccine. Whether AIDS, which spread from diseased old world monkeys can be likewise defeated, remains to be seen.

Somehow scientists failed to connect this same adaptive principle of Nature to the use of "miracle" chemical pes-

---

* First demonstrated in 1796 by England's Edward Jenner in the prevention of smallpox.

ticides. Consequently, in only 60 years we have created 600 known species resistant to chemical controls. The solution? create new pesticides.

The faster a pest species can reproduce, the less time it takes for it to become immune. Microbes reproduce at 20 minutes of age, humans at about 15 years. That is why our pests will always be ahead of us in coping with each new pesticide. An added worry is that some of the more persistent pesticides are building up in the environment and increasingly harming both humans and ecosystems. Nature's continuing biological magnification of these persistent pesticides and hormones is wreaking genetic and other havoc in wildlife species. And they are now known to be a growing factor in human degeneration.

b) Pest Control Without Chemicals

In 1900 practically no chemical pesticides or fertilizers were used. None are allowed in producing certified organic products today. Instead, organic growers use human labour, imported off-farm nutrients (e.g. seaweeds), and an increasing array of tricks. The fossil fuel input however, compares with conventional farming (20 calories per calorie eaten). Among their very successful techniques for conserving nutrients while controlling weeds, insects and diseases are:

- Consistent sequences of <u>crop rotations</u> (include cover crops), which feed the soil, not the crop. This greatly reduces the various soil losses from erosion, leaching and crop removal—all major soil destroyers in modern agriculture.

- **Mechanical pest reduction** by reducing the pest's niches, hand picking, using pest resistant plant varieties (like scab resistant Nova Mac apples); excluding insects via various row crop covers (like Re-may); flame weeding, sprinkling salt on cabbages to control cabbage worm, or ashes to deter slug, and earwig traps.

- **Control of animal pests, product quality and parasites** by proper breed selection, careful feeding, uncrowded range rearing and sanitation practices concerning food and water containers, holding facilities and manure handling.

- Since **killing pest animals** often increases their reproduction, it is better to fence them out, or cover crops with netting, remove their preferred food and cover habitats or use scaring devices like mechanical exploders.

c) Pest Control With Chemicals

(i) Background

Our 60-year history of increasing dependence on man-made chemicals to control pests has worked miracles— for the short term. But more and more of those chemicals no longer work because many pest species have developed tolerance – 600 so far and counting. Meanwhile, out in the environment, many of our earlier pesticides are being "biologically magnified" to toxic levels, which consequently affect humans.

Traditionally, short-term solution has been to develop new chemicals. But a new pesticide now costs millions of dollars to develop and register for "safe use".

Another answer to the chemical resistance problem is to select and breed "hardier" crops like Roundup-Ready soybeans. In this case hardier refers, not to the pests, but to the plant's tolerance to registered pesticides (from whose makers nearly all of the research money comes).

Besides not addressing the problem of escalating pesticide contamination, these new genetically modified crops are now cross-pollinating with the major weeds and making them resistant to herbicides—super weeds impossible to control by existing means.

As to the effects of Genetically Modified Organisms on long-term human health, the scientific community is divided. Europe has banned GMOs. "Organic" growing of non-GMO canola in Saskatchewan is now contaminated by GMO pollen, which excludes it from the European market. Since a similar scenario for GMO wheat (already being planted) would mean a major export loss for Saskatchewan and Canada, this case was before the Supreme Court of Canada in 2003 and was not allowed. Genetically engineered crops are increasing and now cover 20% of croplands worldwide. They not only contaminate the soils where they are grown, but their pollen is carried far and wide by wind and water, contaminating susceptible plants far from sources of application.

(ii) Mighty Mites Still Mightier

Varoa mites became established in Florida honeybees in the 1920s. They now infest practically all colonies in North America. Without chemical controls, mites will destroy a hive of bees within two years.

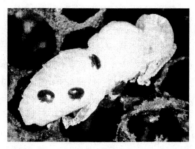

Infested brood
Mites live on honey bee larvae

Although the chemical pesticide Apistan and several other pesticides have been successful in controlling bee mites, resistance to Apistan is spreading rapidly. It soon proves useless wherever used exclusively by even a few beekeepers. Chemical resistance can be <u>delayed</u> only if <u>every</u> beekeeper alternates pesticides regularly (i.e. uses Apistan in the fall and Formic acid or other acids in the spring). Once Apistan resistance becomes universal, it is useless in the rotation. Mite antibodies simply treat it as part of the environment to which they have become adapted.

Since eliminating Varoa mites is now impossible, the industry urgently needs a non-chemical, non-resistance-creating way to keep mites in check. Queen breeding for resistance seems to be one possibility, but a very long-term project. A 2001 Spanish experiment using mineral oil (www.beesource.com) showed some promise and deserves further work, since mites cannot achieve resistance to oil (as apple growers well know from its use on mites for 100 years).

The goal is to reduce the size of the mite's niche in the hive. Killing mites by chemical poisons can only be a short-term solution. This is a fundamental principle of Nature.

(iii) Other Introduced Species

- ### Coyotes in Newfoundland

A predator niche was left open in the 1800s when 45 kg (100 lb) wolves were eliminated over the large settled areas of North America and reduced to "threatened" status even in some northern wilderness areas. In the early 1900s the little 12 kg (25 lb) gopher-eating prairie coyote of the American Midwest crossbred with one or more wolf sub-species, creating what we now know well as the purebred eastern coyote averaging about 18 kg (40 lb)

While the original wolf needed huge territories for surviving, and depended on large prey such as moose, caribou and deer, the smaller and more versatile eastern coyote has evolved much wider eating habits in smaller territories. So much so, that it easily lives almost invisibly next to human habitation. Individuals even learn to eat dog food out of an outside dog dish and not be seen by the dog's owner. As a result, eastern coyotes are now entrenched in all of the eastern North American mainland. In the 1980s they recently crossed the Strait of Belle Isle and/or Cabot Strait and are now sweeping across Newfoundland. While snowshoe hares (themselves introduced) will no doubt be their staple food, they are expected to play havoc on the caribou calving grounds.

As with our exotic mite, getting rid of coyotes is impossible. This has been amply demonstrated by a century of shooting, bounties, poisons, traps, and snares. On the contrary, it has improved both its productivity, survivability and wilyness beyond compare. However, populations left to Nature, coyotes and prey will eventually stabilize. Before this happens with Varoa mites however, they will have killed off the many wild honeybee (Apis) populations.

- ## Beaver Boom-Bust Phenomenon

When I began my study of beavers in Nova Scotia in 1965, people were witnessing the "bust" and blamed over-trapping. Biological specimens collected from trappers indicated otherwise. In fact, harvesting "to little too late" had allowed the beavers to over-populate and to destroy the best trees in their habitats. And in Nova Scotia beavers have no serious non-human predators. Over most of the western region counties, over-browsing, cutting and flooding had lowered the habitat's carrying capacity, hence fewer beaver.

In 1908 in response to a long-time absence of beavers over all of the province except in one remote western Tobeatic region (due mostly to over-trapping), the government prohibited the taking of beavers in Nova Scotia. The first open season and bag limit was in 1943. By then beavers had re-occupied all of the western counties and were causing a lot of damage by flooding roads and cutting trees.

In the eastern counties beavers had arrived much later. As a result their populations were still in lush virgin habitat and thus booming. So the bag limits were liberalized there to take advantage of their very high kit production.

The combination of intensive forest cutting practices in the late 1800s and early 1900s creating ideal early succession trees such as aspen and red maple for beavers to spread into. Moreover they enjoyed complete protection and were encouraged through live trapping and release into vacant areas.

These eastern county beavers, just reaching their boom stage, was evidenced by their optimal reproduction, survival and growth. There, adult females—over three years old— averaged six kits every year; all three-year-olds, and even some two-year-olds were pregnant . Adults were huge, with a few pelts measuring 92" (234 cm combined length + width). About 50% of the trapped harvest was made up of kits (i.e. six-months-old beavers averaging 7 kg (15 lbs).

In spite of generous bag limits to keep populations in check, all biological indicators declined and natural mortality slowly increased. So harvesting declined too. Our management goal was to keep populations stable by adjusting harvests for optimal use of this natural resource over the long term. An added consideration was that beaver colonies foster many other wildlife species, including fish.

It should be clear from this example why "boom-busts" occur. Each wildlife species has its own limiting factors that keep it in balance with its physical habitat and with all other interacting species, both plant and animal. This is a fundamental law of wildlife ecology.

Can we apply this insight to human populations? One big difference is that domestic animal and human habitats are artificially maintained. Another is that our food carrying capacity has been expanding in the West thanks to artificial energy. This has allowed human populations to expand. But our boom is now being maintained not only by total dependence on oil and other artificial energy sources, but by exploiting the human and natural resources of the "under-developed" 80% of the planet. Their hunger and privation stand in stark contrast to our western affluence. Food for thought…..

An Honest View of "pecking order" Hierarchy in Humans

Have you ever wondered why American tourists tend to be treated differently than Canadians in certain foreign countries? The "pecking order" phenomenon of hierarchy may have first been scientifically studied in chickens. Basically, each individual hen in a flock of say 200 served by one dominant rooster, instinctively gets to know every individual "above" and "below" her, and behaves accordingly in her daily activities. Only the odd squabble between two hens will break their very peaceful routines. Similarly, every tribe of apes in the jungle takes care to establish "territory boundaries". These minimize conflicts and save energy.

Paranoid forest-based apes on Danakil Island, forced to shift onto a strange habitat, had to learn cooperative new ways of social life. This helped to avoid constant tribal battles both on land and in the water, on the wide-open but narrow beaches where they spent much of their time (Elaine Morgan). When these proto-humans left Danakil and began their hunter-gatherer life-style on the African steppes, one major conflict-avoidance tactic between tribes was to keep widely dispersing into new territory. This also provided more pristine habitat opportunities, a strategy used to this day. The settlement of North America first by Neolithic or Palaeolithic Asians and later by Europeans, is a case in point. Globally speaking, Celt-led nations are still in their "boom" stage mentally and developmentally.

Brief History of Human Wealth, Power and Hierarchy

Webster's dictionary defines wealth thus: "All material objects having economic human value in existence at one time." To over simplify, the concept of wealth likely began when the pebble tool led to killing more prey than immediate needs dictated. Previously, apes foraged for food individually as needed. Any surplus (wealth) would have been shared by the mate or tribe or stolen by others. Such accumulations of food-wealth conferred a real survival advantage i.e., less time spent on risky food gathering. But it created the need to defend this wealth against marauders. "He who has cattle on the hill will not sleep easy." (Irish proverb).

Wealth as a status symbol was first expressed by hierarchy in defence of mates. The invention and use of superior weapons to create chiefdoms and kingdoms was the next phase in the evolution of wealth of civilizations.

Next came competition for control over "underdeveloped areas", leading to the colonial phase. The past 200 years of industrial use of fossil energy (coal, oil and gas) to fuel massive production of goods and services, as well as weapons, ended in WWII with the U.S. emerging as the global "alpha" nation.

Today it is economic might plus superior weapons which determine the pecking order of nations. Meanwhile terrorism, while not new, has reached unprecedented massive and far-reaching levels that are ultimately impossible to defend against.

Part III

The Nature of Celts

Overview: Why Celts?

Did you ever wonder why we Celt-led nations, especially the Group of Seven industrialized nations (G-7s),
have so much more wealth than other nations? Why a
G-7 citizen's life is deemed by us, to be somehow more
valuable than that of an individual from the teeming
hordes of poor foreigners? (Notice the CBC always
reports the number of <u>Canadians</u> killed in any disaster
overseas.)

When I began to read about Celts on a recent trip to
Ireland, I began to recognize some of the predominant
characteristics in myself and in those around me. Of
course "Celts" are well mixed with a wide diversity of
other cultures of the human family, their offspring creating the expected blends of characteristics. Nonetheless,
here, without fear or favour, are my blunt, empirical
conclusions and deductions about what I see as most
significant:

Today's currently wealthiest nations are Celt-led and

dominated; the exception is Japan, which however, adopted Celt tactics in the 19th century.

The wealthy G-7 nations produce, use and trade "goods and services" for profit, globally. This gives them a great advantage.

An objective assessment of the history of this wealth, including its effect on people and Nature, never mind its oxymoronic unsustainability, leads to some sobering inescapable observations and deductions. In my opinion, they are leading us all to catastrophe. Fortunately, the most prosperous nations which have the most to lose, also have the greatest potential for change, plus the wealth to bring it about. We need only be convinced of our peril.

The above conclusions have been disclosed on the internet, without fear or favor, for all the world to see. But none of them is discussed publicly (let alone addressed) by common people of Celt-led nations. Without an emergency about-face, future generations of Celt-led people face a bleak existence. They are blissfully unaware that the era of cheap oil is over and what this will mean for our comfortable life-style.

Celt History

Celts came out of the north sometime in the late Stone Age 4000 years ago, and migrated into central Germany. Here they farmed and left most of the oldest Celtic artifacts we find. By then Celts were already culturally highly advanced, having their own language, social development, arts and crafts. That they possessed exceptional artistic talents is attested to by their intricately crafted golden ornaments, such as torts.

Fearsomely aggressive young Celt warriors spread south through the Roman Empire as far as present-day Greece, pillaging and raping as they went. Some settled there. Celts never took slaves and never became slaves, choosing suicide instead. They readily inter-bred with any culture, producing the wide diversity of talents we see today in the most wealthy and prosperous nations, which now dominate global resources, trade and development.

Human traits That Celt-led Nations Excel In

Simplistically, the modern Celt traits that are largely responsible for their descendant's wealth and power today boil down to a polished talent for killing, thieving, lying, cheating and hypocrisy.

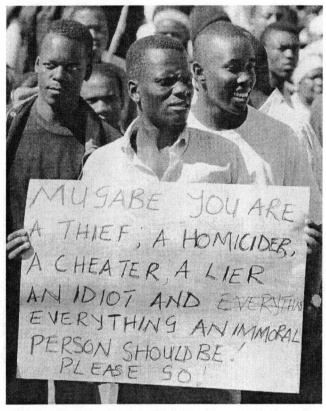

In the past three weeks, 10 supporterss of Zimbabwe's opposition have been arrested, and several beated or tortured. Among them was an oppoosition MP, who says: "This regime has lost control of its senses."
Dominating traits of Celt-led nations: Thieves, killers, cheaters, liars and hypocritical bigots.

Following WW II, which was the "stage battle" between two Celt cultures, the United States, under president Harry Truman, committed America to developing what we call the Third World, or underdeveloped nations of the world. With this stroke of the pen, 80% of the world's population was declared "poor" and in need of help to develop their resources. From this action, followed the enslavement, impoverishment and ever starvation seen in these countries today. Many of these countries had already been colonized by other wealthy nations

Ancient Origins of These Celt Traits

We were never really meant to be predators. For millions of years we had evolved like our shy jungle ape relatives of today. The aquatic ape hypothesis provides a vivid and realistic account of the possible origins of many of our unique human traits, and of how each of these might have become entrenched during a period of long evolutionary adaptation. Elaine Morgan used known evidence and facts to rationally construct her *Aquatic Ape Theory* to account for hitherto largely unexplained anomalies between apes and humans. Her arguments are compelling.

On an individual level, I have tried to assemble known stepping stones to account for my own personal feelings, and to better understand the empirical observations and conclusions discussed below. While certainly not defini-tive, they give food for thought.

My Empirical Conclusions

Humans are mammals, and like all mammals have nev-er stopped being part of Nature. We have never stopped having territory and establishing cooperative hierarchi-cal pecking orders. These still serve original functions of peacekeeping. As always, our body language during day-to-day encounters, expresses our pecking order, usually instantly, to everyone within range, though we often use words to hide it. These basic human traits and purposes extend to nations and to the balance of power, and are accepted by many individuals in each nation. Human history is largely the history of warfare, still ongoing and, paradoxically, still vital to peace.

While humans are part-time predators (by definition we depend partly on killing prey), unlike full-time predators we have only a weakly developed instinct against murdering our competitors. Instead, humans have "morals" established by religion and law. Courts (tribunals etc.) establish the guilt of violators and impose penalties. This creates liars, cheaters and hypocrites. In contrast, all animals are amoral. Domestic dogs can be taught to feel "ashamed", and will exhibit this by body language that is usually distinct from fear. A dog may cheat by stealing food, but, once having eaten it, won't lie (i.e.: pretend he didn't).

Similarities of Apes and Humans

Humans and apes have a very strong affinity and kinship, although some humans instinctively deny this. Physically, our genetic make-up is almost identical to that of chimpanzees. We share an essentially cooperative nature and a great many other characteristics—e.g. facial expressions—are clearly evident to any serious observer. We both thrive in warm climates, eating various warm-season fruits and vegetables. We are very sociable, loving and lovable. Unlike true predators, apes and humans react with great horror to killing and death in general. In contrast, a cat will catch a mouse, then enjoy playing with it before finally eating it. A wolf will eat a live deer without evident emotions, while the helpless animal looks on in shock. The sport of bullfighting, still popular in Mexico and Brazil, re-enacts the killing of large prey.

Behaviourally, humans and apes are sexually polygamous by nature. Extended families come naturally also, along with necessary hierarchy, honesty, and ethics in small groups. Both humans and apes have strong "clan" territory instincts. Neither has strong instincts about either sanitation or the use of commons. For example, water is used for sewage, even their own drinking water.

Humans differ from "cavemen" only in that Celt-led nations developed hygiene (Greek for health), and have supposedly given all humans "rights". We've gone overboard with both. Hygiene only truly emerged after Pasteur proved that germs cause disease and that these germs were spread by contact, mostly by both animal and human excrement contaminating drinking water and food. Today, in spite of daily showers, soap and automatic washers and dryers, our bodies and our sewers are toxic chemical sinks, contributing to our current plague of ill health.

In the Middle Ages neither humans nor animals had <u>any</u> rights. Humans could be sentenced to life imprisonment without ever being charged. Many were flogged with 100 lashes as an example, and their lifeless body then turned over to relatives for burial. Since animals weren't considered human, their screams of pain were sometimes used for entertainment. "Habeas Corpus" supposedly gave humans some rights, yet torture is still rampant globally. However Germany and some American States have recently conferred legal status on animals.

Some Human Anomalies

1. Killing

One of our early anomalies was diet. As noted, our first ancestors had to learn a radically new way of gathering food, predator style, from their new, predominantly salt-water habitat—and learn it evolutionarily fast or die out. Fortunately, this brain—enhancing high phosphorus diet enabled a long series of social changes that continue to the present day. Undeniably they have made us biological predators. But we exhibit some very divergent lapses. Some of these we hide for shame, while others we disturbingly accept as being perfectly normal.

For example, a part of us abhors killing, yet—where food, defence and emotions are concerned we accept it as essential. On the other hand our "emotional logic" (instincts) override some distinctly un-predator-like behaviour, such as the unconscionable, no-limit killing of strangers, enemies and pests that seem to impede our whims. "Kill it. If it's good for a scoff, eat it. If it ain't, kill it anyway. It ain't fit to live." Or consider the motto inscribed on a U.S. peace soldier's eagle medallion, worn around the neck: "Kill first; ask second.". Finally, unlike apes, humans can and do live on a vast diversity of edibles, the latest thing being the largely artificial array of processed food and drink.

2. Emotional Extremes

Humans have great imaginations, perhaps evolved in connection with speech and storytelling to enhance survival. Such imaginative extremes may share roots with extremes of fear, prejudice, compassion, hate, cruelty, greed. Our passion for leisure and sports activities may have similar roots. One of our oddest attractions is the primitive sport of boxing. The drama of this fairly bloodless self-defence fascinates young and old alike.

3. Change Aptitude

Since the beginning really, the survival of hominids depended on change. From the invention and improvement of the pebble tool to the vast and expanding diversity of our 78,000 chartable "causes" (many totally contradictory) we relish change. Doing good, improving or just creating change come quite instinctively to humans. In fact, without change we stagnate both individually and culturally.

Any species living in Nature changes very slowly because under normal conditions habitats also change very slowly. But in today's artificial zoo-like conditions, human change is rapid and rampant. "Rule #1: If we don't look after you, someone else will" (seen in a Customer Service).

In today's human world, quite unlike the situation with apes, the principal changes and shifts in hierarchical dominance are initiated in temperate zones notably by Celts. Their ancestors learned to live in the shadow of the glaciers—literally at the cutting edge of constant and extreme changes. So Celts in wealthy countries drive global change through relentless innovation, whether in clothing or nuclear bombs.

4. Aquatic Tendencies

As noted under Evolution Theory (Part II), humans exhibit a surprising number of distinctly un-ape-like human peculiarities closely resembling those of mammals fully adapted to life in the sea. It was these which in the 1940s led Sir Allister Hardy to his original hypothesis and to publish "Was Man More Aquatic in the Past"?

Morgan describes each of the following exquisitely:

- Our body fat is attached to our skin, as in seals, whales and walruses

- Our bodies appear almost hairless except on the head

- Human body hair and nose shape are streamlined for immersion

- Humans have tears for excreting excess salt (apes don't)

- Our babies can swim long before they can walk and, indeed, seem to prefer being born under water

- Our commonest sexual posture is belly to belly as in rotund aquatic mammals (male apes mate from behind only).

- Humans have a very close affinity to bodies of

water, especially the sea

- Humans crave salty foods, indeed, seafood is preferred worldwide, and even as few as one meal of fish per week offers very measurable health benefits

- Humans, like whales, have a diving reflex which automatically slows the heart rate during immersion

- Pronounced webbing between fingers and thumbs in 6-10% of all human populations.

5. Sexual and Reproductive Anomalies

Perhaps one of the most "un-natural" anomalies of humans is that copulation takes place regularly and irrespective of the oestrus cycle of the female—even throughout her pregnancy. Apes and all other mammals copulate only at the time the female is receptive—once per cycle. (Only occasional exceptions have been observed in wild apes.)

Other human sexual anomalies include:

- Human females tend to accept sex on demand by their mates, in some cases almost daily. Human males need sex at least once a week and may masturbate to get it. Apes in captivity may masturbate, but apes in the wild don't need to, and rarely do.

- Female humans tend to display their numerous sexual attractants socially, even when not consciously looking for a (usually monogamous) mate. Females tend to select a single male based on looks, sociability, security and his capability to provide for them long term; often some children aren't "his" (10%). Old saying: "It's a wise child that knows its father".

Elaine Morgan's *The Evolution of the Child* (1995) discusses in detail various other human anomalies related to infancy and childhood. Some of these include:

• Babies at birth are relatively large and fat, with a waterproof skin and the ability to swim.

• A prolonged period of babyhood followed by dependence on family for 15 years or more.

All these traits are totally unlike those of apes, yet each must have conferred powerful selective advantages on humans over thousands of years.

6. Diet (animal protein)

As already indicated, the ancient need for ordinary apes to eat seafood to survive was likely the catalyst of our human evolutionary journey. When after many thousands of years, the predatory behaviour became entrenched, seafood was largely replaced with land animal protein gleaned on the African steppes, where most early hominid evidence has been identified.

In spite of very valid reasons for vegetarians refusing to eat most meat and fish (they do need some animal products), there are very basic reasons (besides health) why humans should eat seafood and meat. And those reasons may be our strongest and best motivation for greatly improving current animal protein sources, including safeguarding purity and not destroying the environments which produce this protein .

Nearly three quarters of the globe is covered with water. Most of this water is in the oceans, which are capable of producing vast amounts of good human food. It is perhaps cheaper to harvest this than any other high quality foods. But our oceans are heedlessly polluted and mistreated in many ways. Also, Celt-led harvesting methods discard large by-catches and destroy essential fish habitat on the ocean floor.

Much of the land near where people live is un-tillable and only suited to the grazing and browsing of animals that humans have survived on for millennia.

The production of non-animal human food requires (i) either a great deal of human energy or (ii) the input of more artificial energy calories than the resulting food contains delivered to the consumer.

"Organic and Free-range" soil management farming is being recognized in wealthy nations as the cheapest, long term system to ensure safe food and water quality, health and food security.

7. Miscellaneous Anomalies

A. "Cooperation"

Ivan Sanderson, (*Living Mammals of the World,* 1955), concludes his section on humans with another anomaly not found in apes: "Men are cooperative except in their own territory or 'nest', where they <u>often</u> display the most independent and aggressive attitudes to other members of their own species".

This uniquely human quirk is no trivial matter. Both women and children live with serious personal threat, often unrecognized—the threat of injury and death, in fact. Examples include the old "rule of thumb", which made it illegal to use a stick bigger than a man's thumb to discipline a wife or children. "Bullying" by groups, and by boys, seems to be a related phenomenon, with both murder and suicide often the end result, as recent news media reports have demonstrated.

Two un-ape-like anomalies of humans
- *Extreme aggression, even in their own families.*
- *Domestication of a predatory animal.*

B. Dogs

One of the most incredible anomalies of all is that of a prey animal (humans) developing a very close domestic relationship with a predatory animal related to wolves— the domestic dog. Dogs, domesticated for more than 15,000 years, have been selected and moulded to serve a wide diversity of important functions down to the present. Traditionally dogs were used as early warning signals around camp (wolves can't bark), as well as for emergency food, pulling sleds, guarding homes, locating game and killing vermin. Modern uses include guiding the blind, tracking game (and humans) and, increasingly, as companion animals for the lonely. A few "work dogs" are still used for herding domestic sheep, for locating and retrieving hunted game and for sled racing by clubs which enjoy the challenge and fresh air.

Many dog stories like "*Lassie Come Home*", "*The Incredible Journey*" and "*Bel Ria*" contain far more truth than fiction about our very close human-dog emotional ties, including their unconditional love.

My own dog experience has been with two setters, and Erica and I now have a Little River Duck Toller. As a working coldwater retriever, she embodies those traits which humans have for millennia, been relentlessly breeding "Dilly's" attentiveness and responsiveness to our idiosyncrasies are truly astonishing. It's as though her only reason for existing is to do our bidding. Our job, as trainers, is to make sure she understands, in dog language, what we can reasonably expect of her.

Her eyes, ears and incredible nose are ever alert to do our bidding, as best she or any dog can. Failure to obey is our fault, not the dogs. Go back to the book, or ask a trusted expert. Dogs, like children, are the responsibility of adults. An untrained dog is a nuisance at best and a liability at worst. Over half of all dogs under two years old have to be put down due to improper care and training. "Brothers and sisters, I bid you beware of giving your heart to a dog to tear." Kipling—*The Power of the Dog*.

The Nature of Celts in Canada

We Canadians find ourselves physically cloistered and secluded, almost strangers from and to the rest of the real world. After 500 years of development more than half of our 30 million people are living in a 200-kilometre corridor along our southern 6000-kilometre border with the United States. The other half is sprinkled very unevenly over one of the biggest and coldest countries in the world.

Canadians have inherited vast resources, many of which have been "creamed", picked over, taking the best and leaving the rest. We live "high on the hog" by exporting our best capital wealth, mostly as raw material, to the United States. America's 300 million people have come to depend on it. Indeed, America's advanced technology and related weaponry would, if need be, defend Canada as a very cooperative, friendly, independent, next-door neighbour, and a major resource warehouse. Canada's minimally equipped Armed Forces are used mainly for peace-keeping missions authorized by the United Nations, or for humanitarian aid (as in the tsunami tragedy of, Dec. 26/04).

Canada is an exceptional, even unimaginable country! It is impossible to grasp the magnitude, beauty and diversity without travelling by land from sea to sea to sea. Flying over the country in daylight may leave a lasting mega-impression, but this gives far too brief a glimpse for the mind to even begin to understand or grasp our incredible inheritance and its great diversity of people.

The Biological State of Canada Now

Let me very briefly summarize my conclusions about the nature and biological course of the dominant Canadian Celts, in our 500-year conquest of Canada. I've tried to be objective. I've tried honestly and realistically (not pessimistically), to restrict myself to things I've seen as a layman plus things I've been trained to see as a biologist practising for more than 25 years. My chart below may help.

The 500 year Celt Colonization of Canada

Century	Population	Development Stages
1500s	Less than a half million	Harvested cod, walrus and whales and developed fur trade with the Natives (who fully occupied Canada).
1600s	One million?	Celt settlement and colonization on the east coast and the St. Lawrence.
1700s	Two million?	Celt pioneered agriculture across Canada, largely displacing Natives.
1800s	Average of three million	Boat building, timber trading and cod on the east coast, grain across the prairies, salmon on the west coast.
1900s	5-30 million	World trader of wood, grain, fish and minerals; hydro and nuclear electricity.
2000	30-60 million?	Global raw resource provider – 90% to the U.S. Chief prosperity from wood, seafood (farmed increasingly), mineral, electric power, water. Tourism a major service resource and economic engine.

While some of my conclusions may be over-simplified, I leave them for bona fide experts and lay people to judge, challenge, or improve on.

Since about 1900, Canada's steadily rising per capita use of, and now increasing dependence on, artificial energy such as oil, has been paralleled and related to destruction of natural resources, and pollution, by chemicals especially, of all waters, air, soil and food. The degeneration of Canadian health has been worsened, not only by these harmful environments but also inactive life-styles plus poor eating habits and unhealthy junk foods in particular.

I attribute the roots of above destruction to ignorance of our own basic Celt nature and/or leadership, and to our evolutionary development. Failure to grasp or understand the severity and extent of this damage, inflicted by even commonly accepted practices, and consequences on current and future Canadian generations is a green issue.

The machinery Canadians use today to extract resource products—fish, farm crops, lumber and pulpwood, minerals etc.—is so massive that even their operators easily overlook the great harm being done today and down the road.

Shift in Canadian Concepts and Practices Concerning Killing

Relevant Underlying Facts

Approximately 80% of Canadians are now urban dwellers, and most rural dwellers are urbanized too. That is few of us obtain her or his basic needs personally from Nature anymore. Before 1900 most of us lived a rural life directly involved in farming, fishing, trapping, forestry, and so on. Even town dwellers had some direct contact with their country cousins, if only to take their kids to "the farm" occasionally. Few do so today, and even if they did, little or none of what they would see would in any way "connect" with any <u>basic</u> need at home.

The highly developed division of labour between town and country offers great advantages, except in one crucial dimension—Nature. The more and the longer people are insulated from their dependence on Nature, the less understanding of it they have. Indeed, this is approaching zero in 99% of modern Canadians. One symptom of this disconnect is that the concepts of "Killing" are increasingly divorced from biological realities. Psychologically speaking, there is a disconnect. Author Richard Louv goes further calling it Nature Deficit Disorder (*Last Child in the Woods*, 2005).

Some Examples of Our Nature-Disconnect in Canada

Most of us know of and accept the killing of domestic animals for food—so long as we don't have to kill them ourselves. Vegetarians though, have very valid practical and ethical reasons for rejecting factory-farming models.

Similarly, most of us accept modern large scale farming practices—even though, as discussed in Part 1, it necessarily involves applications of ever-increasing tonnages of diverse chemicals to fertilize and protect crops against rapidly evolving pests including weeds, insects, fungi and bacteria.

By far the most devastating killing practices are being perpetrated directly and indirectly by the developed world in "under-developed" areas of the world for profit.

In Nature by comparison, all killing is solely for food, and part of a long-term ecological process. In this process humans were for eons a prey for large carnivores. Recently, a hiker, killed by a bear, was shown on the internet partly eaten, still wearing a running shoe. More recently, a hiker was attacked and wounded by a wolf. "Law of tooth and claw" has vivid meaning for us.

Not surprisingly, it is usually the people who are the most disconnected from Nature (while protesting the opposite) who are the most vociferous in their objections to "sport hunting". It so happens that this is the only kind of Nature-approved, purposeful killing. And those people who hunt, fish and trap—even ritualistically once a year —gain a memorable first-hand experience in Nature, as well as good food to eat.

Groups of these "consumptive" users of wildlife—hunters, trappers and anglers—have globally always been in the forefront, of movements to preserve and conserve wild land resources. Helped by environmental groups, their funded lobbies have helped Canada set aside vast blocks of land that would otherwise not have been preserved. Generations to come, will better appreciate and understand the true meaning of Nature in its broadest sense.

Like other Celt-led nations, Canadians want to continue living "the good life" outside Nature. But our continued wilful destruction of Nature's capital regressing Gaia to earlier stages is madness.

We need to stop trying to "save the Earth" (Gaia is a Big Girl and can save herself.) Ask instead "How can we save ourselves"? Do we need more proof of our Nature-disconnect, humanity's nemesis? Certainly Gaia doesn't need humanity.

For example, what happens when China's 1.3 billion people follow our lead? Beijing is said to register 30,000 new cars every month. (The Nova Scotian, Aug. 13/06)

Part IV:

Evolution of Human Health

Good Health Based on Evolution

1) Background

In most wealthy nations with a few exceptions, the current dreaded causes of death and disability—cancers, Alzheimer's Syndrome, heart disease etc.— -were rare or non-existent before 1900, even in the elderly who did expect some arthritis. But why should degenerative diseases kill two-thirds of us now?

Modern health professionals, whose stated goal is to remedy this unhealthy state of affairs, seem to be as much in confusion and disagreement about the causes and cures as people were during the Black Death plagues of 500 years ago. Certainly the average person doesn't know.

Our conventional medical practitioners are trained to diagnose and deal with germ-caused diseases. Their incredible diagnostic, chemicals and surgical technologies, can usually assess the immediate cause of any malady very accurately. But they tend to treat everything—even mental syndromes and symptoms—with

---

antibiotics, hormones, chemicals, radiation or surgery. Even though most physicians would agree that most of our poor health begins with our unhealthy eating habits, plus inadequate daily exercise, they can do very little to change their patient's behaviour. By contrast, alternative practitioners, rather than just treating symptoms in isolation from conditions, rely on non-invasive, holistic and preventative treatments well documented in health literature and widely available.

An estimated four million wealthy Canadians are so convinced that toxins and chemicals are making them ill that they are willing to pay for treatments by these alternative health providers. Many more take vitamins, minerals and other supplements, and buy steadily increasing quantities of bottled water. Consumption of certified organic food and other health products is increasing at an average annual rate of 20%.

Even so, official government recognition of environmental and dietary factors as health determinants could hasten improved regulation, enforcement, subsidies, education and taxation, in mainstream medicine. These in turn would hasten needed changes and provide better health for all of us,—not just the wealthy.

2) Keeping Healthy

Around 1900, when our current degenerative decline can be said to have begun, most Canadians were living and working on farms, and in lumbering or fishing villages. Horses, and oxen, not steam, gas or electricity provided most of the motive power. For all practical purposes, there were no electric lights, no tractors, no cars, no planes, no television, no computers, no radios and no telephones. There were no "chemical" fertilizers or pesticides. People ate real health food, mostly fresh and in season, that had been "organically grown" before the phrase was invented. Apples, root vegetables and cabbage lasted all winter in home cellars. Fruit, vegetables and pickles were all home-preserved with sugar, salt and vinegar, or by drying. Big game meat was commonly preserved in jars.

People lived very active lives year round. A railroad worker was given half a day off on Christmas Day. Kids walked to school, church and stores, and played softball, soccer and hockey using willow sticks and frozen horse balls, all for the fun of it. Most kids had chores to do before and after school.

Pollution? What pollution? (In those days the dictionary definition was "emission of semen...other than in coitus".) No garbage either. Water? Lots of pure spring well water carried in buckets. Artificial food—i.e. foreign to our northern evolution—was pretty well limited to grains, white flour and white sugar. Their ill-effects were easily offset by otherwise good diets and by constant physical activity.

Before 1900 there were no electric lights to keep us up during long winter nights. We spent long hours asleep, or at least were much less active than during the long busy days of spring and summer. (Studies indicate that the blind have less cancer and suggest that we need darkness to restore our "over-illuminated" bodies and minds.)

3) Chaotic Transition Between Plagues

Even the scientific community was understandably slow to recognize the enormous implications of Pasteur's 1880s discovery linking microbes with specific diseases. As Arno Karlan so beautifully describes in his book *"Man and Microbes"* (pages 134-147), these were extremely chaotic times for the developed countries. The terror of hundreds of thousands dying horrific deaths by cholera and other water-borne diseases caused panic and suspicion. People thought they were being deliberately poisoned. It took many years for people to accept the germ theory and to undertake the massive, costly measures to keep sewage from entering water supplies. One hundred years later we had Ontario's Walkerton municipal water pollution, which sickened some 2,000 residents and killed 7, to show us how critical this is.

Medical doctors of these times were hated, shunned and killed, for their punitive practices were often worse then the disease itself. Hospitals became known as places in which to die, not as places of healing.

In 1936, when I was 6 years old, I remember seeing my grandfather Dr. Robert Hillary. He died of Alzheimer's the same year. He was the town physician in Aurora, Ontario, and it was said that there wasn't a house he hadn't visited. His residence is now a museum, complete with his brass sign at the door,

"ADVICE GRATIS DAILY from 10 to 11.

The magic of his day lay in teaching patients and town officials the principles of sanitation. Equally important was diagnosing germ diseases and preventing the spread of germs to others. Previously whole families could be wiped out in two weeks.

4) Crippling Choices

Modern wealthy Celt-led nations are run by science, their faith based on more than 100 years of magic bullets, which seem never to have failed us. Today nothing seems to be true unless it bears the stamp of a published scientific authority.

Celt-led nations have all but defeated northern germ diseases. At the same time, oil-based chemical fertilizers and pesticides have largely if temporarily, delivered us from the natural limitations of food production. Scientific medical discoveries continue to dazzle us with cures, just as antibiotics, discovered in 1928 and developed during WWII dazzled our forebears. Antibiotics are now a mainstay of agriculture as well as of medicine. Likewise immunization, drugs, surgery, CAT scans, x-rays, MRI and radiation are all used to keep us safe, even as our graph of crippling afflictions climbs steadily upward.

Now, a new twist, just in time, Genetically Modified Organisms, (GMO), are made possible by splicing genes from totally different species to create desired features in otherwise ordinary organisms, This supposedly leads to better foods and medicines, more manageable crops, and so on.

Science is now seriously holding out to wealthy investors the promise of fulfilling every dream of perfection imaginable: an end to every physical defect including ageing; greater food production for the poor and starving, improved nutrition from both plants and animals, thus solving our increasing nutritional deficiencies, perfect health forever. At the same time, promoters promise environmental friendliness. What more could we ask? A gene for sanity?

5) Soils and Health

What is soil productivity? How and why does soil affect your health today?

A   Overview

In 1925 Russians published the first, now universally accepted, system of scientifically classifying all soils. They pioneered the realization that soils are a very dynamic living entity whose vital physical and chemical characteristics reflect the dominant vegetation which produced them over the past 10,000 years since glaciers retreated. The United States governments' geological soil classifications had to be all done over.

a) Climax Vegetation

The so called climax vegetation is a self-perpetuating complex of biodiversity which naturally develops on various soil. Sites are delineated by soil type, slope, aspect (e.g. south facing vs. north facing). Soil types are shaped by the following physical characteristics, which in turn favour a particular climax vegetation (e.g. Acadian forest, salt marsh, peat bog, meadow etc.)

* Climate – average seasonal precipitation, temperature, etc.

* Soil texture – from pure clay to coarse gravels and rock.

* Soil drainage – from swamps to well drained gravel.

* Mineral and chemical content of source material.— called parent material

b) Parent Material

This refers to the raw mineral material which underlies any developed soil. It was left on the surface, usually on top of bedrock, after the glaciers melted between 8 and 12 thousand years ago.. Many different organisms reworked this material to create the original soils now found in only a few undisturbed areas after 10,000 years of changes.

Soils have visible layers called horizons (see figure below). Each layer has distinct characteristics that affect overall fertility and health, depending on how we treat it.

A Simple Soil Horizon of a Forest Soil

O horizon: Organic matter

A horizon: Rich in organic matter

E horizon: Zone of Leaching (if present)

B horizon: Zone of accumulation

C horizon: Partially altered parent material often oxidized by weathering

R horizon: Unaltered parent material As deposited by glaciers (till—glacial drift—or in old lakebeds

Bedrock

(Figure 4.1 P. 52 of Edwad A Keller's *Environmental Geology* 1988)
A simple soil horizon of forest soil
Prairie soil depths are deeper in high grass prairie (3—4' Manitoba) and shallower in dryer areas (12" Alberta)

B. *Soil Fertility*

A soil's innate productivity is determined by its Cation Exchange Capacity (CEC). This is its ability to hold and exchange macro and micro chemicals and elements with plant rootlets. Clay and humus have the greatest CEC; granitic sand and rock have the least. Seemingly by magic, mixtures of plants in successive sequences over millennia created living soils from chaotically scraped materials left by the glaciers, either as drift or sediments.

Thus in western Manitoba and eastern Saskatchewan's "tall grass prairie", ancient lakebed prairie grasses created soils more then a metre deep. Over time, in terms of value per hectare to wildlife and humans, these are the world's most productive soils. In eastern North America, woodland-created soils tend to be innately less productive per hectare. This is due to greater nutrient leaching by greater precipitation, but also to higher acidity (lower PH). Trees, especially conifers, acidify soil more than grasses do, and acids reduce the intake of nutrients. These two biomes, namely grassland and woodland, originally created most of today's agricultural soils in temperate areas.

Human crop production, plus removal of trees and animals, over time tend to slowly remove and/or destroy some of the original nutrients, fertility and productivity. Moreover, plowing and cultivating to aerate the soil increases oxidation and burns up organic matter. This releases nutrients, thereby increasing productivity for a year or two: but thereafter yields decline rapidly unless lost nutrients are replaced by manures, NPK fertilizers, or compost.

In addition to macronutrient losses (NPK), micronutrients vital to human health (some not necessarily important to plant health), also decline, reducing the value of food the crops produced. This applies for both the domestic animals and for humans depending on them.

A local Wolfville dairy farmer, Herman Mentic, was plagued with calcium deficiency in his milk cows. Even when calcium was added to the daily rations, cows regularly developed milk fever paralysis after calving. These problems vanished when he stopped using commercial NPK fertilizers and used composted manure instead. It took him four years to convert.

In France, a serious copper deficiency causing cancer was solved by adding copper to the soil where the milk cows grazed. (Soil, Grass and Cancer; André Voisin Acres Books 1999.) This classic text links health of animals and humans to the mineral balances of the soil. It was translated from the French with copyright © 1999 by Acres U.S.A. ISBN: 0-911311-645

In stark contrast to human degrading of soil fertility by removing crops from the land and not replenishing it, Nature continually builds soil fertility through selective plant succession. Thus any natural area not seriously disrupted by human activity (or temporarily by Nature's regulating "disasters"), always tends toward another stable climax vegetation.

Indeed, Nature has conveniently arranged for natural disaster, most commonly by fire, to periodically set back any given succession for a time, before returning it to the climax condition in the end. It seems ironic that the black spruce staple of Newfoundland's forestry industry is dependent on forest fires for its existence. Modern fire prevention, followed by clear cutting followed by tree planting does, mimic fire somewhat by suddenly removing most of the humus capital. But, unlike fire, it depletes minerals and trace elements which the harvested wood contained.

Animals living in Nature remove nothing permanently from the soil or its fertility. In fact, like the plant species in any biome, the corresponding animal species are essential to Gaia's soil-building process. It's only when this natural wealth is "stolen" from the land and soil nutrients not replaced, that soil deteriorates.

C. *Farming And Forestry Losses from Conventional Practices.*

Soils all over the world are being steadily impoverished with each crop harvested whenever plant nutrients are not being replaced adequately. Conventional agriculture replaces macro plant nutrients by costly applications of NPK manufactured from non-renewable fossil fuels. Independent research indicates conclusively the continuing loss and impoverishment of our agricultural soils, especially heavily cropped soils. (See 2002 publications, Kentville Agriculture Centre, N.S.)

Forest soils are almost never fertilized, so each successive harvest conclusively shows its downward nutrient spiral. (P. Ogden, Dalhousie, Studies in Forest Decline in N.S.) A study in Newfoundland, designed to evaluate the effects of conventional NPK fertilizer on a black spruce stand, showed a doubling of yield. But the amount of usable fibre, to make newsprint, remained unchanged. (R. van Nostrand, St. John's personal communication). (It would however, increase the amount of building material such as stud wood.)

D. _Where Do Lost Soils Go?_

a) Losses from Cultivated Crop Lands

Chemical fertilizers destroy or alter soil organisms, which normally conserve and build soil fertility. The abnormal releases of natural nutrients, plus the added artificial ones, are both "loose nutrients" free to be leached into the ground water and or to wash into surface streams. The result is pollution, over-fertilizing and clogging with vegetation all the way down to the sea. (Chesapeake Bay, USA, was one extreme example, requiring compulsory watershed-wide soil conservation practices to alleviate the problem.)

b) Losses from Non-Cultivated Crop Lands

Plant nutrient losses attend each removal of any material from any soil. (Studies done on cattle-grazed meadows indicate significant nutrient losses by erosion also.) Nature's nutrient replenishment comes from a minimum of 20 Kg. of nitrates per hectare per year in precipitation. In addition, miniscule but significant amounts of minerals are added from lower layers of parent material by deep-rooted plants like alfalfa, and on marine terraces from marram grasses whose roots may go down 20 metres or more.

c) Losses Due to Development

Vast soil acreages are displaced for buildings, roads, parking lots, mines and many other ongoing and increasing developments.

E. _Deposition of Soil Nutrients Harvested_

a) Consumables

Feed for domestic animals is converted into animal products like meat, milk, eggs, wool and leather, all of which leave the farm. By-products left on the farm such as straw and manure, are best worked back into the soil to replenish at least some of the loss. But significant amounts are permanently lost in the products, and must be somehow replaced in order to keep up production.

Manure from humans eating the soil nutrients in their food is flushed into the sewers and may or may not be subsequently "treated" before the nutrients find their way to the sea. In either case they will be mixed with the wide diversity of chemicals used in "cleaners". These also go down the drains not only of households, but also of hospitals, stores, businesses, universities, research institutions and governments. Then there are pharmaceutical growth enhancers etc. Another contaminant is glycol antifreeze leaked by millions of ailing moor vehicles and flushed from airport runways after being used for de-icing.

b) Non-Consumables

Forest products are used for fuel, fibre (paper, etc.) and building materials (lumber, etc.). Paper products, which dominate Canada's forest output, can be recycled, composted or burned, and the ashes put back on the land.

c) Landfills

Whatever "garbage" we can't recycle, compost or burn, we transport many kilometres to large and small landfill sites to be buried under soil material. Leachate from these sites slowly enters the water systems for many years, causing costly environmental problems for future generations.

This good fertilizer is mixed with cleaners in our sewers.

F. *Domestic Animal Evolution – pets to PETA*

a) *Basic Stages*

Basic human compassion for animals, even as we killed them for food, no doubt led first to our rearing orphaned baby wolves, horses, cattle, sheep, and such. Eventually realizing the benefits of this, we become more sedentary as game became scarcer. "Dogs" were trained to hunt game and haul sleds. Cows and sheep were used for milk, wool, meat, and so on. Jean Auel's stories (see Selected Literature) places this stage at the end of the last Ice Age.

Once we in the north temperate region became sedentary, we learned to fence out marauders and to house and feed livestock over the winter. This meant greater food security. This phase lasted into and beyond the 20th Century, even as animal husbandry was transformed from a family or clan endeavour to larger community operations of the type I was trained for.

Today "factory farming" of livestock in giant facilities holding, say, 1,000 milk cows, 10,000 pigs or 100,000 chickens provides most of our livestock needs. While most operations are still family-based, now they rely on high-tech, state-of-the-art, automated machinery and buildings. A single family-owned unit of thousands of hectares can now be operated by a few technically and chemically trained workers and ever-larger machines.

b) *Modern Concerns*

In addition to the concerns I raised under "soil fertility", our mechanically efficient factory livestock production creates other major concerns. Always these are humaneness and excess manure.

Humaneness

PETA (People for Ethical Treatment of Animals) is most concerned for the welfare and humane treatment of all the domestic animals entrusted to their keepers. This concern extends not only from our pet cats and dogs to the cramped living quarters of livestock. These lead to unhealthy, conditions requiring periodic (if not constant) feeding of antibiotics and/or hormones; mutilation (e.g. de-beaking poultry to prevent cannibalism); ammonia-caused pneumonia in poorly ventilated pig facilities, and the use of "research guinea pigs" like rats, mice, rabbits, cats, dogs and anthropoid apes in cruel experiments for many un-essential purposes to ensure human safety concerning new chemicals, cosmetics and the like.

Manure

The massive scale of these factory farms creates a correspondingly massive problem of manure disposal. For instance, surplus pig manure leads not only to an incredible stink when spread on the land, but atrophies nearby streams, killing the fish by depleting oxygen. Proper composting and aeration would not only solve both problems, but would conserve enough nutrients to pay for it.

c)   *Questions Concerning Ethics and "Quality" of Canada's Animal Products*

Canadian consumers have, year round access to an abundance and variety of the top quality animal and other food products, unexcelled in the world. These top quality, health-giving foods cost relatively little compared to both our take-home pay (only 10%) and to the actual hidden cost of production (already detailed under "Soils and Health").

So, what could mainstream Canadian consumers possibly object to, as vegetarians do in increasing numbers? Here are some of the major concerns—and not only of vegetarians by any means:

- That animal fats are unhealthy and are making people fat (both proven to be false).

- That raising animals under crowded factory farming conditions is unethical. Also that unnecessary amounts of grain, some of it suitable for human food is fed to animals for meat and dairy production, instead of being sent to starving Third World people. (Such simplistic notions and actions would not be practical).

- That in the process there are antibiotics, hormones and pesticides in raising factory-farmed animals is unsafe in that animal food products may retain traces of these, and/or may be contaminated with E. coli or salmonella pathogens.

- That the fish species in greatest demand – e.g. northern cod—have been decimated and that others, like haddock, are now being over-harvested. Also that the process, huge by-catches of unprofitable species (often 25%) or over-quota catches which are dumped overboard, dead.

- That fish harvesting practices using large "draggers" destroy ocean floor habitats, thus jeopardizing future fisheries.

G) *Summary and Conclusions*

Major Schemes Providing Human Food and Fibre:

a) Conventional Farming and Forestry practices, using gigantic tractors and fossil fuels, are destroying the productivity of living soils whose innate fertility took millennia to create. These practices also undermine ecosystems and extirpate species at increasing rates. And soil impoverishment exacerbates our increasing health degeneration.

b) Third World Methods. Most of these people still depend on proven small-scale soil nutrient recycling practices (including use of human manure), which have sustained their soil integrity and productivity for 4,000 years.

c) "Organic and Free-range" Soil management farming is being recognized in forward-thinking, wealthy nations as the cheapest, most sustainable system for maintaining water quality, healthy soils and food security.

d) Seafood Cheap prices, especially for dragger-harvested species, depend on subsidizing too many draggers, which causes large-scale destruction of ocean-bottom fish habitat. (A civil case is before the Supreme Court of Canada to stop this practice). Such market-driven harvests not only over-fish preferred species, but waste very large by-catches and over-quota harvests, through dumping dead fish.

e) Harvesting Wildlife Resources. Most of us in wealthy nations, living outside Nature as we do, may not deem wildlife resources a valid source of our future food and fibre. (Instead we are accustomed to meeting all our needs at the Mall.) Many, raised on Hollywood renditions of Nature abhor the very thought of killing and eating wild creatures. Yet humans have been feeding and clothing themselves that way for millennia. As a result, large groups of non-government organizations (NGOs) are basically opposed to killing any wildlife, especially not whales, elephants, seals and other "species du jour" being promoted to raise do-good money.

Scientific management of wildlife resources depends on maintaining healthy habitats for each species. Overall environmental quality must first be assessed, then vigilantly maintained. Simply leaving Nature alone and free of human involvement, to manage any wildlife (as some very powerful NGOs advocate), is naïve at best and at worst, implies a deceitful hidden agenda having little to do with conservation (or even the organizations' stated or implied objective).

Wildlife resources must be valued as a trade commodity—because they are. As such, they are and will continue to be managed wisely by people closely dependent on them. Removing those incentives from outside pressures not linked to sustainability is false conservation hiding behind popular rhetoric

All Nature is threatened by human population expansion

and other pressures, that are becoming global in scope. Usually wildlife will survive in developed nations in proportion to its value to the people using it. Cancelling this use not only deprives its rightful users, who in most cases have been managing it sustainably, but also deprive the community as a whole: for example, seal hunters on Canada's east coast.

There are <u>many</u> cultures, both primitive and advanced, which still carry on traditional wildlife uses and who derive many benefits from this. Some wealthy nation NGOs, are deliberately blocking sound scientific management of wildlife globally via the Convention on International Trade in Endangered Species (CITES). Eugene Lapoint headed CITES for 20 years and disagrees with this. *"Embracing the Earth's Wild Resources* (2003) documents his observations and clearly indicates, not only the importance of these resources but also the major impediments to their conservation.

Clearly, harvesting wildlife resources <u>is</u> already a major source of human food and fibre. Its sustainability <u>is</u> vital to humans and the Earth. Such resources must be managed using the best possible scientific data. In this, the prime users are usually the first line of defence. Non-use is no longer an option (if it ever was). Certainly "no-kill" NGOs, and/or nations using them to hide behind racial prejudices, must no longer be tolerated at International meetings as they have been in the past (<u>*Embracing the Earth's Wild Resources*</u>).

Canada has immense wildlife resources of inestimable value both domestically and for trade. Unfortunately, salt-water fisheries harvesting methods as we have seen, are unsustainable. In contrast land based wildlife resources are very well managed. This is because they are regulated using biological data respected by primary users (a process still not followed by our saltwater harvesters).

Recent events have begun to harm not only Canada's very successful domestic biological management traditions, and, for reasons just discussed, international wildlife resources as well. Very wealthy anti-kill NGOs, garnering money from many trusting naïve supporters via very high quality, deceitful propaganda, are now blackmailing companies and governments. For example PETA (People for the Ethical Treatment of Animals) through very persuasive images and literature, are overtly promoting non-use of any animal for <u>any</u> purpose, including as companion pets. Young supporters aren't told this, of course.

Canada's vast wildlife heritage is there for all—present and future generations—to use sustainably forever. Consumptive use of wildlife by natives and pioneers has been universal—vital since the beginning. Its tradition is equally important today across Canada. Except for a few species decimated early on, furbearing animals are more abundant than ever before. Game animal and predator harvests are very well regulated. The animals are being skilfully hunted, trapped and shot in the age-old pursuit of recreation as well as for the healthful, low cholesterol meat, not to mention fur and the trophies and indelible memories.

There is no logically sound reason to remove these frontline, long-term largest contributors to conservation of wildlife. "Saving" Nature via non-use is the cruellest myth. Humanity's future depends on wild resource preservation through understanding and wise use.

H) *Consequences of Canada's "cheap food policy"*

a) Almost entire dependence on fossil fuels for food production, distribution and export.

b) Destruction and increasing pollution of soils and environments.

c) Increasingly cheaper food imports threaten local farming security.

d) Animal products from factory farms may be more at risk from contaminants of bacteria, antibiotics, hormones or pesticides. Humane issues are also a growing concern.

e) The basic management directions of modern mega-farms cannot readily be altered to accommodate factors beyond the farmers control, such as world trade wars, imports and weather. The very survival of many farms is in the hands of governments.

Part V:

Commons

_____

Common Celt Concepts of Commons

In the beginning, our earthly human Garden of Eden included the whole earth—the ultimate commons. All species in it were and are subject to one irresistible biological commandment: "Thou shalt strive for overall ecological stability".

The advent of urban sprawl and agricultural expansion resulted in what Garret Harden called *The Tragedy of the Commons* (1960). His thesis was that abuse of the Commons (i.e. water as in pollution; pastures as in overgrazing and so on), continues because it always benefits the individual violator, whereas, everyone else suffers only a little. Over time, however, the resource itself becomes depleted.

Globally today there are so many degraded commons – all waters, all environments the atmosphere—that, as in times of plague, imminent multiple catastrophes force us all to recognize and prioritize. We need solutions to such problems as species extinctions, global warming, rising rates of human diseases including degenerative ones. We need to curb not only human development, but our addiction to massive and escalating use of fossil fuels, which we linked to climate change, or at least to exacerbating natural long-term trends.

The penalty for violating Nature's evolutionary stable strategy (ESS) is extinction. Shape up or die. This applies to humans too, of course.

What is our Celt strategy now? What changes in strategy are possible at this late stage of realization? For, incredibly, in a couple of million years we have gone from a shy hominid to a global force able to inflict global scars. Among these consequences we will have to learn to live with are depleted ozone layer, nuclear radiation, degraded and destroyed natural resources, toxic persistent pesticides, garbage pollution, dependence on life-support systems based on fossil fuels without which civilization would quickly come to an end.

Part of our current Celt predicament stems from too narrow a concept of excellence. This ideal Celt concept which we all strive for and live by, fails to honestly count the real cost. Unless we come to grips with the fallacies which support this illusion and weigh the now familiar consequences our excellences will have been achieved in vain. Our losses will outweigh our pitiful gain.

In a word, we must return to more holistic concepts of our current global Commons. Celt-led nations, leading the world's people, must and can change. The World Wide Web could be both catalyst and clearinghouse during the short time we have left before total chaos brings down our "house of cards."

This treatise can really only serve as a "seed", not even as a handbook, to enable Canadians to tackle our share of the unfolding catastrophe, and to apply our inherited advantage of vigour and versatility. By doing so we can perhaps avoid some of the major pitfalls ahead. Simply grasping the incredible opportunities most Canadians enjoy in this land of plenty should spur us on.

Our purpose and wish is that you , whether Celt, a hybrid or wannabe, do what you can your way. We G-7s have no choice. We are far in the lead, and will continue to lead.

The outcome is up to us, and only us. Good luck. We'll need a lot of it.

Selected Literature

Books

Allman, William F. *The Stone Age Present.* Touchstone Books 1994

Appleman, Philip. *An Essay on the Principle of Population.* By Thomas Malthus. W. W. Norton Co. New York, London. 1976

Atlantic Geosciences Society, *The Last Billion Years.* Nimbus Pub. Ltd., Halifax N.S. 2001

Ardrey, Robert. *The Territorial Imperative.* Delta Books 1996

Atkins, Dr. Robert C. *Health Revolution* Bantam Books 1990
Dr. Atkins Vita-Nutrient Solution. Simon & Schuster NY, N.Y.

Auel, Jean M. *The Clan of the Cave Bear.* Random House 1980
The Valley of Horses. Bantam Books1982
The Mammoth Hunters. Bantam Books 1985
The Plains of Passage. Bantam Books 1990
The Shelters of Stone. Crown Publishers N.Y. 2002

Bott, Robert with David Brooks and John Robertson. *Friends of the Earth* 1983

Browne, Dr. Joy *It's a Jungle out there, Jane.* Crown Publishers 1999
The Nine Fantasies That Will Ruin Your Life. Three Rivers Press, Random House Inc.1998

Cahill, Thomas. *How the Irish Saved Civilization.* Random House Inc.1996 Toronto

Collier, James. *The Hypocritical American.* MacFadden

Books 1964

Copp, Terry (with Bill McAndrew) *Battle Exhaustion*. McGill-Queen's Press 1990

Courtney, Bryce. *The Power of One*. McArthur & Co. Toronto 1999

Cruise, David and Alison Griffith. *On South Mountain*. Penguin Books 1998

Cudmore, L. L. Larison. *A Natural History of the Cell*. Times Books, N.Y. 1977

Cuthbertson, Brain. *Wolfville andGránd Pré Past and Present*. Formac Publishing Co. Ltd. 1996

Dale, Ann. (in collaboration with S. B. Hill) *At the Edge*. UBC Press 2001

Dale, Ronald J. *The Invasion Of Canada*. James Lorimer 2001

Daly, Martin and Margo Wilson. *Homicide*. Aldine deGruyter 1988

Darwin, Charles. *The Origin of Species*. Modern Library 1859 *The Descent of Man*. Modern Library 1871

Davidson, Heather Ann. *Civvy Street*. Lancelot Press 1996 Hantsport NS

Dawkins, Richard. *The Selfish Gene*. Oxford Press 1976

Deer, Thomas. *The Terror That Terrorism Produces*. Discussion Discourse Autumn 2002

Diamond, Jared. *Guns, Germs, and Steel*. Norton Books 1997 *Collapse*. Viking 2005

Dodds, Donald G. *Nova Scotia's Snowshoe Hare*. Dept. of Lands and Forests, 1987.

Eaton, S. Boyd. *The Palaeolithic Prescription*. Harper & Row. 1988

Fitzgerald, Edward. *The Rubáiyát of Omar Khayyam*. Avon Books 1941

Fogle, Bruce. *Dog Training*. DK Publishing Inc. 2002

Frankl, Victor E. *Man's Search for Meaning*. Simon & Schuster Inc. 1230 Ave. of America N.Y. N.Y.1984

Fukuoka, Massanobu. *The One-Straw Revolution*. Rodale

Press Inc. 1978

Gallant, Roy A. *The peopling of Planet Earth*. Macmillan Publishing. 1990

Gilbert, Frederick F. PhD. And Donald G. Dodds, PhD. *The Philosophy and Practice of Wildlife Management* Krieger Publishing Co. 2001

Goodall, Jane. *In the Shadow of Man*. Houghton Mifflin 1971

Harding, Garrett. *The Tragedy of the Commons*. Armonk 1968

Heinberg, Richard. The Party's Over. New Society Publishers, Gabriola Isl. B. C. 2005

Herm, Gerhard. *The Celts*. St. Martin's Press Inc. 1975

Herman, Arthur. *How the Scots Invented the Modern World*. Three Rivers Press N.Y. 2001

Herscovici, Alan. *Second Nature*. CBC Enterprises 1985

Hines, Sherman. *Outhouses Address Book*. Firefly books Ltd. Willowdale ON. 1996

Huggan, Isabel. *The Elizabeth Stories*. Harper Collins Pub. Ltd 1990

Karlen, Arno. *Man and Microbes*. Touchstone Books 1996

King, F.H. *Farmers of Forty Centuries*. Rodale Press 1911

Klein, Naomi. *No Logo*. Vintage Canada Edition. 2001

Fences and Windows. Vintage Canada Edition 2002

Kramer, Mark. *Three Farms*. Atlantic Monthly Books 1977

Lawless, Charles. *And God Cried*. J. G. Press 1994

LeGuerer, Annick. *Scent*. Translated from French by Richard Miller, Kodansha Internet. 1992

Leopold, Aldo. *A Sand County Almanac*. Ballantine books 1949

Livingston, John. *Rogue Primate*. Key Porter Books 1994

Lott, John R. Jr. *More Guns Less Crime*. U of Chicago Press 1998

Lovelock, James. *The Ages of Gaia*. 1988

Gaia: a New Look at Life on Earth. Oxford Press 2000 (originally pub.1979)

MacCrossan, Tadhg. *Druids* Llewellyn Pub. 1995

McGhee, Robert. *Ancient People of the Arctic.* Canadian Museum of Civilization. 1996

McMahon, Anges. *Celtic Way of Life.* O'Brien Educational 1976.

McHughen, Alan. *Pandora's Picnic Basket.* Oxford Press 2000

Miner, Jack. *Jack Miner and the Birds.* Ryerson Press 1923

Morgan, Elaine. *The Descent of Woman.* Souvenir Press 1972

The Aquatic Ape. Stain and Day Books 1982

The Scars of Evolution. Oxford Press 1990

The Descent of the Child. Oxford Press 1995

The Aquatic Ape Hypothesis. Souvenir Press 1997

Morgan, Marlow. *Mutant Message Down Under.* Harper Collins 1994

Morris, Desman. *The Naked Ape.* Bantam Books 1967

The Human Zoo. Bantam Books 1969

Mowat, Farley. *People of the Deer* 1980

Rescue The Earth. McClelland & Stewart Books 1990

Nova Scotia Department of Lands & Forests. *Notes on Nova Scotia Wildlife.* 1980

O'Donnell, James. *Celtic Proverbs.* Appletree Press 1996

O'Griofa, Mairtin. *Irish Folk Wisdom.* Sterling Publishing. 1993

Orchard, David *"The Fight for Canada".* Multimedia Publishing Inc. 330-4999 St. Catharine St. West, Westmount, QC H3Z 1T3. 1998

Orwell, George. *The War Broadcasts.* edited by W. J. West, 1985 BBC Publication by Gerald Duckworth And Co. Ltd. ISBN 0563-203-277 (BBC) The Complete Novels, Eric Blair Estates 1976, 1983.

Peter Dr. Laurence J. and Raymond Hull. *The Peter Principle.* McLeod Limited 1969 Toronto

Potter, Charles Frances. *The Faiths That Men Live By.* Ace Books Inc. 23 West 47th St. New York, NY 1955

Priesnitz, Wendy. *Challenging Assumptions in Education.*

The Alternate Press 2000.

Raftery, Joseph. *The Celt.* Mercier Press 1964

Rampa, T. Lobsang. *The Third Eye* Gorgi Books, Transworld Pub. 1966

Ratey, John J. (with Catherine Johnson) *Shadow Syndromes* Bantam Books 1998.

Ray, Dixie Lee. *Trashing The Planet.* First Harper Perennial 1992

Ross, Sally with Alphonse Deveau. *The Acadians of Nova Scotia.* Nimbus Pub. 1992

Sanderson, Ivan T. *Living Mammals of the World.* Hanover House 1955

Saunders, Gary L. *So Much Weather!* Nimbus Publishing 2002

Seton, Ernest Thompson. *Two Little Savages.* Dover Publications 1903

Schaef, Anne Wilson When Society Becomes An Addict. Harper & Row 1987

Schumacher, E.F. *Small is Beautiful.* Abacus Books 1974

Shaw, Matthew. *Great Scots.* Heartland Associates Inc. 2003

Shepard, Paul. *The Tender Carnivore and The Sacred Game.* Scribner's Books 1973

Nature and Madness. U. Of Georgia Press 1982

Shiva, Vandana. *Stolen Harvest.* South End Books 2000

Smith, Margaret *Ritual Abuse* Harper Press San Francisco 1993

Smith, Melissa Diane. *Going Against the Grain.* Contemporary Books, a div. of McGraw Hill Co. 2002

Suzuki, David. (And Holly Dressel). *From Naked Man to Superspecies.* Stoddart Pub.1999

Swift, Jonathan. *Gulliver's Travels.* ISBN 0-19-519978-2 1977

Sykes. Brian. *The Seven Daughters of Eve.* Norton and Co. 2001

Towers, Julie. *Wildlife of Nova Scotia.* Nibus Publishing1980

Vallée Brian. *Life With Billy.* Seal Books Toronto 1986

Van Gelder, Richard G. *Animals & Man Past, Present, Future.* F.E.E Inc. 1972

Voisin, André. *Soil, Grass and Cancer.* Acres Books 1999

Walters, Richard A. *Gun Dog* Clarke Irwin Ltd. Toronto 1961

Watchtower Bible and Tract Society. *Mankind's Search for God.* New York 2001

Wiley T.S. (with Brent Formby, Ph.D.) *Lights Out.* Pocket Books 2002

Williams, Redford, M.D & Virginia, PH.D. *Anger Kills.* Harper Perennial 1983

Wilson, David M. *The Vikings and Their Origins.* Thames and Hudson Ltd 1970 London

Wilson, Edward O. *On Human Nature.* Harvard University Press 1978

The Future of Life. Random House Canada 2002.

Wylie, Philip. *The Magic Animal.* Doubleday Books 1968

Wright, Ronald. *A Short History of Progress.* House of Anansi Press Inc. 2004

Zinsser, Hans. *Rats, Lice and History.* Black Dog & Leventhal Books 1934

Other References and Personal Correspondence.

- *A Living Machine* 1996 © Copyright: New Internationalist provided by Tim Upham, Inspector Specialist, Nova Scotia Dept. of the Environment & Labour. (902) 679-6086 *Analysis of Heavy Metals in Sewage Sludge, Sewages and Final Effluent.*
- *Alive.* Publisher Seigfried Gursche.
- *Canada and the Human Environment.* Environment Canada. 1972
- *Clean Ocean Action – the end of sewage sludge dumping* 1996.
- *Dumping Sewage Sludge on Organic Farms? Why USDA Should Just Say No.* 1998

- *Future Agriculture – Man in a Natural Ecosystem.* Agrologist 1975
- *New Internationalist.* New Internationalist Publications. 2002
- *"Public Legal Education, Now".* "Legal Information Society of Nova Scotia".
- *State of the World, 2001.* Publisher W. W. Norton & Company. 2001
- *The Tragedy of the Commons.* Garrett Hardin. Science 162 (3859) 1243-1248 1968
- *What If It's All Been a Big Fat Lie?* Gary Taubes 2002
- *World Watch.* Publisher Lester R Brown. 2002
- *Yesterday in Acadia* Acadian Historical Village 1987
- *20s Eugenics Plan Horrifies...*Globe & Mail, Dec. 7/02 (E17)
- *Discourse & Disclosure,* Winter Spring, Sue Potvin PO Box 1112 Greenwood NS B0P 1N0 2004
- *The End of Cheap Oil* National Geographic June 2004
- *Bulletin Scan.* Elizabeth May, Sierra Club of Canada. 2003
- *Agriculture in Kings Co., Real values and Real progress.* Jennifer Scott. Et. Al. 2000
- *Adbusters Media Foundation,* 1243, 7th Ave. W. Vancouver BC V6H 9Z9

Some Relevant WEB Sites.

Hubbert Peak of Oil Production. www.hubbertpeak.com

Will The End of Oil mean The End of Amarica? 1/6/2005 www. Commondream.org

Life After the Oil Crash. 1/6/2005 www.lifeaftertheoilcrash.net

Hormones in shampoos and early puberty: www.eurekalert. org

Environment Canada – Endocrine-disrupting substances in the environment www.ec.gc.ca

Scripps Howard News Service: Chemicals and ill health: www. ec.gc.ca

Appendix

I first thought of writing this treatise while visiting Ireland, my mother Hillary's roots. Then, reading Celt history, it suddenly hit me "My God, I'm descended from these Celts.

The archetypical Celt was described as pale-skinned, tall, blond-haired, and blue-eyed. Because of religious belief, they were unafraid to die. Roman scribes tell of Celt warriors going into battle stark naked except for spear and shield. Roman soldiers ran away as these savage giant strangers descended upon them, whooping and hollering, long blond hair standing on end or blowing in the wind.

Celt history indicates only that they came "out of the north in the late Stone Age", complete with their unique language, agriculture and religion. They settled in what is now central Germany where many early artifacts are found.

History details how Celts swept south through the heart of the Roman Empire. They plundered and raped as they went as far as Greece. Celt genetic traits combine well with other races, the descendents exhibiting infinite combinations seen today throughout the world. The wealthy G-7's are run by Celts whose people are the acknowledged leaders of the globe.

Much of the modern G-7 wealth originated as booty taken during many aggression wars. This piracy continues today in the developing world unable to resist our Celtish pressure to develop their resources and to use their cheap labour to feed our ever-growing endless appetite for cheap energy, goods and services.

With almost unlimiting cheap artificial energy Celtish practices are being utilized to extract any and all natural resources. This is going on despite extreme, long-term, often doing irreversible, ecological damage to Global Nature (Gaia, earth goddess), not to mention our own rapidly degrading home environments. The latter degeneration, deteriorating our health, especially in children, who, being the most susceptible to the effects of a sick environment, act as "canaries".

Modern Effluent and Garbage – Waste

"Waste" in Nature is vital. A tree's constantly shedding of bark, seeds and leaves provides the essential resources needed for life under the tree. The relatively few nutrients that are exported from a natural forest are eventually used someplace else in Nature. Practically nothing is wasted (not used) in Nature.

Until recently, Third World people likewise wasted relatively little, their "night soil" is still highly valued for fertilizing crops as it has been for thousands of years.

Compare these practices with those of G-7 countries with their massive outputs of effluent and garbage in our "Age of Energy". Household garbage is trucked many kilometres to "land fills". (We fool ourselves that recycling prevents the upward spiral.) Moreover, much of our increasingly long-lived effluent is toxic to humans and wildlife, and will take millennia for Nature to de-toxify.

Ironically, most of our readily decomposable food nutrients from vast urban centres are being laced with sanitation and other chemicals before being flushed into sewers. The end products eutrophy (over-fertilize) rivers and lakes and pollute the vast ground water sources that recharge our wells and municipal watersheds.

"Stay Calm, Be Brave, Wait For The Signs" – (Thomas King's Dead Dog Café)

I admit I'm arrogant; I come by it honestly. "*Perfect Ape...*" is not intended to be another rant (since I'm a guilty party), nor is it meant to be a piece of pessimism or optimism. It could be taken more as a lament. For my generation witnessed,—no, helped to bring on— humanity's current crises. Yet we seem blind, not only to its causes, but to how we can get out of it, so addicted are we to our energy habit.

I really don't have any grand rescue schemes, except the usual one of keeping Celts healthy. And I have tried to retrace the history of one average Celt, myself, with a view to teasing out some lessons for today.

I see most unwell people today as being not only "lost" as to "Why are we here?" but also blinded by the dazzling glare of our unprecedented, but illusory wealth of cheap artificial energy, without which G-7's and civilization would collapse overnight.

One doesn't need to be a biologist to see how lost we really are. Making this single admission would be one giant step for G-7's and for the now all-powerful Celtish people.

Anyone who has been lost in the woods knows that once you can admit to yourself that you don't know where in hell you are going, the cluttered mind immediately clears. A similar German proverb says, "When things move no more, a light comes on from somewhere and shines right there". (Running around in circles keeps this light from coming on). Admitting to being lost is scary. So we often put this off until "all else fails". Some people lost in the woods are found naked. Others, if found alive, become frightened and may run or hide from their rescuers.

There are standard rules for anyone lost in the woods; often they are crucial to survival. There are only five

1.   Sit down.

2.   Make yourself as comfortable as possible.

3.   Check for resources that are immediately available.

4.   Decide your best prospects for getting "un-lost".

5.   Plan a logical course of action.

In one recently reported case, a young man got his hand caught securely under a boulder in the wilderness. After a few days he realized that his best hope of survival was to use his pocket-knife to cut off his hand. To prevent bleeding to death as he walked to safety, he thoughtfully made a tourniquet from his clothing. He survived to tell about it.

As a biologist, it appears to me that Celt-led nations have but a small window of opportunity now to do what evolution, over many thousands of years, has equipped us to do. Our adaptability and resourcefulness brought us to this pass; all we need is motivation.

We Celt-led nations have all the essential resources—energy, knowledge and wealth. How long can we afford to wait, squandering the last remaining cheap energy and continuing to wittingly destroy the only liveable planet in the known universe, that's harboured, if not created life, for perhaps a billion years. Stop; think. Shall we lose a hand today but survive: or wait for more proof and lose an arm and a leg or our lives tomorrow?

Watching the "poor nations" (G-21s) walk out of the 2003 WTO meeting in Mexico, I'm thinking maybe the "Free Trade" in ideas is going to be much more powerful than the back-room deals in "goods and services". After all, it's the "Global Village Commons" that's at stake. (Maybe the waterlillies will cover the earth on that last day).

Many Third World people now see direct ecological connections of "Population Environment Health" (seen painted on one boat hull in Thailand). First World peoples know these three in reverse:

1. Degeneration of our health;

2. Degeneration of our polluted environments, and

3. Provisioning human population using oil energy to extract more and more natural resources, especially capital resources, which will be vitally needed for the survival of any future generations. Many of

these resources belong to Third Worlds, which are often enticed (or black-mailed) into destroying them as we have destroyed ours.

An Open Invitation to Institutes of Higher Learning to Invent and Establish a Gaian* Chair		
A Table Showing Aims and Objectives of Some Major Endeavors with Suggested Gaian Inclusions		
Endeavor Specialty	Narrowest Purpose	Suggested Gaian Ecological Aims and Objectives
Food	To increase cheap food	(1) To use less energy up to consumption (2) To use less non-renewable energy (3) To restore soil quality and stop polluting water
Health	To treat patient symptoms	(1) To reduce incidence of ill-health (2) To educate patients about preventing ill-health
Life Sciences	To harness Nature	(1) To understand and teach how Nature achieves Her Gaian purposes (2) To work with, not against Nature
Ecology	To understand Nature better	(1) To teach major consequences of violations to Nature (2) To research possible corrective measures
Energy	To increase energy production	(1) To replace non-renewable with renewable energy (2) To reduce energy related pollution—production and uses.

Wildlife Management	To monitor key wildlife & recommend action	(1) To teach in all grades, human-wildlife ecology (2) To encourage Nature experiences including consumptive. (3) To encourage Nature reserves near schools
Anthropology	To understand human history	(1) To honestly portray current biological crisis (2) To project serious scenarios for humanity
Proposed Gaian Chair	(1) To list the most critical Gaian Thrusts (2) To categorize origins of thrusts (3) to produce regular reports for all media	(1) To assist establishing ecological aims and objectives in all branches of higher learning (2) To reward positive initiatives (3) To adequately fund highest quality Gaian material to help fill hiatus in current media (4) To search for interdiscipline-cooperation incentives among competitive branches

*inspired by James Lovelock's *The Ages of Gaia*, 1988
And his best seller *Gaia: A New Look at Life on Earth* 1979.

Post Script

There already exists a spate of highly qualified authors who have detailed HOW we western nations are on a dead-end course – now very clearly understood. Some recent books include: *A Short History of Progress; Collapse; Dark Age Ahead; Last Child in the Woods The Party's Over, The Weathermakers, A Thousand Barrels a Second, The Winds of Change* and *The Revenge of Gaia.*

I've taken the liberty as a wildlife biologist to generalize humanity's prospects from my biological point of view. Far from being ridiculous, negative or pessimistic, I see this perspective as an essential first logical step.

Current actions and attitudes indicate we don't have a realistic idea of where we are going or why. Our 100,000 odd causes, many contradictory, indicate to me we are lost.

If one can admit honestly to oneself that we are lost, the cluttered mind suddenly becomes clear. We stop wasting the last of our reserves of energy going around in circles. We sit down; assess our situation realistically; we eventually agree to a logical plan to carry out, step by step, so to speak.

Humanity's current biological characteristics of global populations can be seen as those of an introduced population (like a pest species). We have now reached our maximum expansion into favorable habitats throughout the planet. This is the boom stage of an expanding population. Evidence of this is very high reproduction and recruitment of young.

However, this boom-stage attitude and effect extends for years beyond the need, as it has in many places depending on the widely varying biological factors of history, education, soil, climate, crowdedness, etc.

One very artificial factor, "oil", has pushed total human population numbers far above any eventual capacity of the planet's natural resources to sustain. Simplistically this has been accomplished in the last 100 years significantly, by using this very abundant, cheap energy source to increase food supplies in a wide diversity of ways – from manufacturing fertilizers, pesticides, machinery etc.

Mass production of paper and building materials was similarly accomplished globally. In both cases we have logically invested our capital to cash in Nature's capital resources for our highest "profits".

However, it is now crystal clear that by this process we have actually been <u>destroying</u> our soils, forests and living wild resources, rather then simply using their renewable annual productions. We also realize now that the scale of this destruction is affecting all life and its ability to flourish as it did 100 years ago. The earth's degradation at present rates causes Gaia to regress to earlier stages.

The West's dependence on using vast and increasing

quantities of oil energy in order to maintain our wealth is causing many cancerous blights, and not only at home. The extreme specter of our Midas golden touch is becoming ever clearer. But unlike the fable, we have no cure for the curse of wealth. The following is only a sample among many.

Increasing non-recycled "garbage" waste costs billions of dollars to collect and bury in huge landfills.

Many of the pollutants in our air, water and food continue to increase and will take Nature millennia to detoxify.

Consumption and waste of oil and other non-renewables won't be available for our grandchildren. Degenerative diseases continue to increase and at younger ages, with no specific cause or cure. We have already lost the skills and infrastructure needed to survive without our vast cheap energy—especially in very deficient habitats.

The potentially good news is that our Western nations have proven many times in the past that we have incredible capabilities to do whatever we put our minds to.

We have the energy, power, knowledge and experience, and for a brief time only, could improve our fast diminishing odds. All we seem to lack is a realistic grasp of our plight and what happens if we fail to act in time to radically alter our own value system at home.

Also, we must be motivated by the realization that a great deal of the mayhem we see in the news daily, is a direct result of how we live at home here, consuming blindly, totally ignorant of effects far and near.

Nature-caused disasters stand in stark contrast to our human-caused disasters. The world's people seem of one mind in helping people hit by hurricanes, floods, earthquakes etc.

I could be wrong. I was once before I thought I was, but I wasn't. (I hope I'm wrong this time too).

About the Author

Neil van Nostrand has always loved domestic animals, and wildlife and Nature. Growing up on the family farm in southern Ontario during the Second World War, he had ample opportunity to study all three. Adult life took him to the Ontario Agriculture College, where he earned a Bachelor of Science in Agriculture (agronomy), then a Master of Science in Wildlife Management. In 1958 he moved his small family to Nova Scotia's Annapolis Valley. For 25 years he did management-research on the Province's wild furbearing animals for the then Department of Lands and Forests. Specializing in beavers, he learned valuable insights into population dynamics and the behaviour of mammals under stress in different habitats. These insights apply just as well to our increasingly crowded planet.

Following retirement in 1985, he and his present spouse, Erica Garrett, pursued biodynamic organic farming for nine years, raising livestock, winter storage vegetables and many fruits including apples. In 2006 the Atlantic Canadian Organic Network (ACORN) presented Neil with the Gerrit Loo Award "in recognition of his contribution to organic agriculture."

Neil is still attached to the land, and to his three sons and one daughter, and to their children. His days are filled with walking, running, hunting, fur-trapping, gardening and tending three bee hives. And a Little River Duck Tolling Retriever is the couple's constant companion.

About This Book

Neil, a 76 year old, grassroots wildlife biologist, put this book together over the last 10 years with his spouse Erica. Its central question is, how could *Homo Sapiens* have managed to dominate every habitable space on earth, and now to have reached this moment of self-destructive wealth and power?

Many of us are in great confusion, as if lost in the woods. Where are we all going and why? The gap between rich and poor; growing global pollution; deteriorating ecosystems; increasing species extinctions; massive war machines ruining millions of lives; rapid global heating caused by human activities – the list goes on.

This book provides a biological understanding of what happened and why. Its biological orientation allows the reader to choose a more realistic, more honest re-evaluation of her or his life.

Part I details our rapid shift away from mainly rural lifestyles when most people lived and worked close to Nature, fully aware that it supported them. Today more than 80% of people live urbanely, isolated and ignorant of the global Nature-system supporting their lifestyles by the use of vast amounts of artificial energy.

Part II looks at our deep human origins and why we are the only species lacking a built-in Evolutionary Stable Strategy for sustainability.

Part III suggests how and why the mainly Celt-led wealthy

nations are leading the rest of the world toward a tragically unsustainable Western lifestyle.

Part IV covers our health history from African origins through plagues, science's defeat of many germ-born diseases and the West's modern degenerative disease epidemic.

Part V alludes to the Global Commons destruction leading inevitably to a global crash, already evident in some crowded African nations today.

The author hopes the book will be a "seed" to motivate people in the Western nations to cool their prodigal lifestyle, for the sake of both present and future generation's physical and mental well being and for the salvation of Earth's rapidly collapsing ecosystems.

ISBN 142511098-3